D0482187

Cloning the Buddha

Advance Praise for Richard Heinberg's *Cloning the Buddha: The Moral Impact of Biotechnology*

"Nuclear energy taught us that the consequences of powerful new technologies are not foreseeable. Can we assess genetic engineering in a context of ethics and the long term well-being of humanity? Richard Heinberg offers us clear, well-balanced information, expert perspectives and opinions, deep questions – everything we need to bring our personal expertise to the ongoing debate. His masterful and sensitive treatment of a very hot issue deserves the widest possible audience."

—Elisabet Sahtouris, Ph.D.,
evolution biologist and author of *A Walk Through Time* and *Biology Revisioned*

"Part philosopher, part journalist, part bodhisattva, Richard Heinberg takes on a labyrinth of subjects: history, philosophy, ethics, and biotechnology. The result is a nimble exploration and an offering to the effort to democratize science."

—Chellis Glendinning, Ph.D.,
author of *My Name is Chellis and I'm in Recovery from Western Civilization*

"This scholarly analysis of the moral impact of genetic engineering biotechnology reveals the kind of science and state of mind driving this new "life science" industry. The evident threats to our humanity and to the future of Earth's creation, which cannot be denied, should concern us all."

—Dr. Michael W. Fox,
senior scholar, bioethics, Humane Society of the United States and author of *The Boundless Circle*

"I would recommend [this book] as required reading in any course on contemporary issues in science, ethics, economics, governance, medicine, or religion. It is written with understanding of the issue, with grace of presentation, and with thoughtful judgement."

—Thomas Berry,
author of *The Dream of the Earth*

Cloning the Buddha

The Moral Impact of Biotechnology

RICHARD HEINBERG

A publication supported by
THE KERN FOUNDATION

Wheaton, Illinois ♦ Chennai (Madras), India

First Quest Edition 1999

The Theosophical Publishing House
P. O. Box 270
Wheaton, IL 60189-0270

A publication of the Theosophical Publishing House, a department of the Theosophical Society in America

LIBRARY OF CONGRESS CATALOGING-IN-PUBLICATION DATA

Heinberg, Richard.
Cloning the Buddha: the moral impact of biotechnology / Richard Heinberg. — 1st Quest ed.
p. cm.
Includes bibliographical references and index.
ISBN 0-8356-0772-0
1. Biotechnology—Moral and ethical aspects. I. Title.
TP248.2.H445 1999
179'.1—dc21
99-19371
CIP

4 3 2 1 * 99 00 01 02 03 04 05

Printed in the United States of America

Contents

Foreword

Dorion Sagan and Lynn Margulis

RICHARD HEINBERG'S BOOK *Cloning the Buddha: The Moral Impact of Biotechnology* focuses our gaze with unique clarity on the fascinating contemporary issues of cloning and biotechnology.

In our view, cloning and biotechnology are not the unmitigated blessing of a gleaming Epcot tomorrow promised by insufficiently accountable corporate interests, whose underlying profit motive raises a cloud of doubt above the scientific research they fund. Nor are they a Promethean horror show of unparalleled dimensions threatening global ecology even as they unravel the moral threads from which life's value has been religiously stitched. Richard Heinberg, who seems to agree with us, makes us aware that what we are already doing transcends the novelist's imagination. He reports objectively on the latest research—epochally symbolized by the 1997 cloning of the first mammal, the sheep named Dolly—and crosses into what some assume must be science fiction, but is not; such as the intentions, announced in 1998, of the appropriately named physicist, Richard Seed, to clone himself, his menopausal wife having agreed to be implanted with the embryo.

Regarding such biotechnological experiments, Heinberg distinguishes between the public's two prevailing responses: "yuk" and "wow." On the yuk side, we have some of the antics of the Monsanto corporation. Its stock price in decline as it markets the genetically engineered bovine growth hormone, the trading public and institutions presumably have looked past the advertising to Monsanto's induction of life span decrease and the disease increase in cows. Also on the yuk side, we have the prospect of genetic equivalents of the grafted and grossly recombined bodies fabricated in H. G. Well's novel, *The Island of Dr. Moreau.* We imagine easily packageable cube-shaped tomatoes, miscreated humans with three hands and scheduled doses of "pass-the-test" intelligence, hybrid interspecies offspring, perpetually growing embryos, and disembodied meat that grows as trimmed steak in restaurant-supply vats. Then on the wow side, we have the prospect of permanent heritable cures for cystic fibrosis, the alleviation of world hunger, and the redesign of ourselves and our loved ones to be healthier, happier, smarter, and undepressed on demand.

To help the reader make sense of this complex mix of blessings, Heinberg brings together and makes accessible a multitude of activities, studies, and observations. His careful and original analysis presents us with the means to think for ourselves.

Consider J. Robert Oppenheimer's reflection on the organization of the multimillion-dollar work that led to the Manhattan Project in Oak Ridge, Tennessee, and the creation of the first atomic bombs and their devastating release on Hiroshima and Nagasaki. When asked why he had advocated the development and use of weapons of mass destruction, Oppenheimer described the solutions to problems of bomb production and used the phrase "technologically sweet." Heinberg rejects the received wisdom that tools and technology are morally neutral but can be used for good or evil purposes. His argument for using political will to restrain research deserves attention, especially from those who, like us, suspect that whatever people *can* do, they *will* do. Yet one thing is clear. The effects of genetic engineering, compared to the bombardment of great urban areas by an atomic or a hydrogen bomb, will never be neatly confined to a single place and time. Through life's tendency to propagate and grow, to incorporate its own past into its present existence, genetically altered organisms can continue to exist and change and spread indefinitely.

Certainly, though, there are ethical issues, as Heinberg highlights: Will the tomato, genomically spliced to include fish genes because of their anti-aging properties, be acceptable fodder for a vegetarian? Can we play wantonly with the very stuff of nature with no hope of predicting consequences that may be everlasting? Is it "fair" for moneyed interests to parlay their wealth into biological advantages as they succeed in assuring good looks, health, and intellect, but only to their privileged offspring? Might not a few members of a biologically-engineered "super-race," by use of proprietous tools that Hitler could only dream about, truly become a Nietzschean *Übermenschen*? Might not these superior humans split off, speciate, and ultimately enslave or extinguish their ancestors, *Homo sapiens sapiens*? Heinberg alerts us to the potential for real-life versions of these science fiction dystopias.

Probably Heinberg's greatest contribution is his independence of thought and his splendid ability to inform us. He makes us aware of how unself-conscious incessant activity leads us to consent to live in a world that, had we thought about it, we never would have chosen. He wants us to see how ignorance and social apathy can generate a biologically impoverished and dehumanized place. For example, Heinberg makes clear that Third Reich directives emerged from the same sorts of scientifically faddish eugenics mind-sets that directed the Rockefeller Foundation and many U.S. governmental "Big Science" ventures.

We agree, yet in our view it is a mistake to interpret genetic engineering as prospectively hideous, or hellishly new on the planet. The "new scientific policies" of Hitler's Third Reich, after all, failed. Mother Nature has never accommodated, and probably never will accommodate, human power fantasies of biological control; she is far too wild. No two beings, even identical twins, are truly identical. Equivalence in identity is a Platonic fantasy. The real world is marked everywhere by difference, impurities, and the vagary and variety of mixture and experience. Genes flow from drug-producing multicellular bacteria to *E. coli* and on to mammals in the blink of a geological eye. Indeed, in science, the second law of thermodynamics tells us that nature displays a probabilistic tendency to mix differences, to break down gradients. Nature hates a vacuum. As microbiologists well know, to maintain a colony—a clone—of a single bacterium for long is very difficult. The palpable world does not promote genetic purity: generation of differences and their dynamic breakdown and assemblage of motley crews of all kinds are not only tolerated but actively encouraged by generous and unpredictable nature.

As we believe Heinberg realizes, community is built into the very fabric of life. We hope we can all agree not to overestimate the power of people. Ultimately, it will be diverse collectives working and playing together—not totalitarian regimes operating under an impossible and ultimately sterile dream of endlessly copied perfection—that will inherit the Earth.

Genetic "engineering" has gone on for eons before humans evolved. The sexual proclivities of bacteria from the beginning of the Archean Eon 3.5 billion years ago until now include a rampant exchange of genes next to which our species' most Bacchanalian orgies are rather subdued affairs. Genes, DNA of indisputably bacterial origin, are found in plants, animals, and fungi; indeed, whole bacterial cells, their genes and all, are absolutely required for all of us to synthesize the vitamin B_{12} we need as we read Heinberg's book. Our genetic identity is incessantly being breached by "foreign" bacterial or viral genes—or whole bacteria and viruses—which are not just invading pathogens but rather are constantly and necessarily being integrated into what we like to think is our pure self. Heinberg appropriately chides our social arrogance and the pathetic assumption that, as a society, we can play God or Buddha or Great Holy Spirit and simply control destiny. He tempers the tendency toward biotechnological alarmism with a dose of reality. But the continuity of the breeder's past with the technologist's present requires repeated emphasis. Every romantic gossip and matchmaker in the history of love has done his or her part to bring certain genes together in one body that might not otherwise have so recombined. Genetic engineering and the ancient "technology" of bacterial sexuality upon which it depends is nothing new. Today's merchandisers and their technologists engineer and profit from bacterial genetic ingenuity: they do not create it.

Heinberg raises genuine concerns. Genetic engineering and cloning present new, fascinating ethical dilemmas for humankind, and Heinberg's thoughtful, enlightening perspective provides a great service in helping us understand the issues. On our part, we can offer hope for the long run; for from the more inclusive and longer evolutionary perspective, we consider it folly to think biotechnology, or indeed, any human-fostered technology, is somehow separate and "above" nature. Nothing people do, we argue, threatens nature as a whole. A thunderstorm contains as much energy as an explosion of a nuclear warhead. A single meteorite impact, such as the Chicxulub which most likely caused a climactic change that led to the Cretaceous extinction that included the death of all dinosaur species, expended more energy than would the simultaneous detonation of the fifty thousand warheads stockpiled at the height of the cold war. We hover, with our millions of propagating progeny, as fruit flies on the hide of an enormous elephant. Nature takes care of herself. And indeed, we would argue, she does so in part by a kind of "biotechnological" play called sex that includes the rampant mixing and matching of (mostly bacterial) genes. Genetic mergers have and will include the natural miscegenation of disease and symbionts. Gene-disseminating tendencies are as ancient as they are fundamental to evolution.

We anticipate lively discussion of the ethical issues Heinberg raises. Readers with very little scientific background, we think, will be awakened to how greatly knowledge has expanded. Yet, as Heinberg so well describes, knowledge—especially technical knowledge—is not wisdom. And it is wisdom he advocates.

Our view is that it is not only physically and practically impossible to stop all sorts of gene manipulation, it is definitionally impossible as well. Sex, symbiosis, and the forming of living communities of different types are all forms of "biotechnology" that not only predate man but have made the evolution of humanity possible in the first place. Cloning as identical copy-making is bound to fail just as it fails naturally and predictably during the course of embryo development. The asexual copies, or clones, of the fertilized egg cell at first resemble each other, but as the body continues to develop these cells differentiate to become distinct lung, skin, brain, and other tissues. The Buddha would not *be* the Buddha if he were cloned. Despite the importance to biology of reproduction, life does not propagate by endless repetition of one or a few platonic types. Ecological succession often begins with a "pioneer" species that rapidly spreads throughout an uncolonized environment. Then a few populations settle in, and their growth slows down as they open the way for new species. These in turn create new niches that, in turn, are settled by other beings. Lush, rich nature at the height of its glory has always used "biotechnology" and has never settled for mere cloning. Rather than try and stop processes that are inevitable and even generative of humanity, we need to have care, precaution, thoughtfulness, and communication about biotechnology's latest—human-fostered—applications.

Heinberg's book serves as a rallying point toward that care, precaution, and communication. It may achieve wide recognition for the thoughtfulness of its statement.

Dorion Sagan and Lynn Margulis

Preface

WHILE THE SCIENCE behind genetic engineering and cloning constitutes a specialized knowledge that only a few people are likely to want to acquire, the basic ethical and spiritual issues surrounding the use of that knowledge can be grasped by anyone willing to spend a few hours reading and thinking. Since our lives are being transformed by these new technologies, I believe we all owe it to ourselves to make that investment.

I began work on this book in January 1998 and spent almost exactly a year reading, interviewing experts in the field, and writing. I brought to the task journalistic skills honed during years of publishing a newsletter on emerging social issues and a solid knowledge base concerning the impact of technologies on human culture and the natural world (already exercised in writing a previous book, *A New Covenant with Nature*, and in teaching college courses on ecology and culture).

It has proven to be an intense year of learning and reflection. I hope that, in synthesizing and interpreting this information, I have made an important subject more accessible to readers.

One thing I want to make clear from the outset is that modern genetic engineering differs profoundly, both in means and potential results, from nature's primordial mixing of genes through sexual reproduction and the action of viruses and other microbes. No amount of breeding or virally induced genetic change will ever produce a stable population of six-legged mammals with the combined characteristics of sheep and rats. But it is possible in principle for human genetic engineers to create such chimeras, because modern techniques can bypass genetic barriers that nature ordinarily keeps tightly in place. With traditional breeding we cannot combine adaptive traits from organisms that are widely separated phylogenetically; but with genetic engineering we can, for example, transfer scorpion genes for venom production to food plants to protect them from insects. One of the potential problems with this is that if the species we engineer has wild relatives—an example might be the sunflower—then it can easily exchange pollen and therefore genes with its wild cousins. The insects that normally control the population of wild sunflowers will be killed by the chemical now being produced not only by the cultivated plants but also by wild populations with which they cross-fertilize. This could be ecologically devastating.

We should care deeply about these effects, because they can only increase the negative impact of human society upon the natural environment. Over time periods that we humans ordinarily care about—our own lives and those of our foreseeable descendants—the potential for such harm is immense. If we continue along our present path, we may alter the world so profoundly that only a small percentage of existing species will survive, and our own kind might well be among the casualties. While the likelihood that nature will recover in some way over the succeeding millions of years may provide a dim sense of comfort, we should do all we can to prevent human damage to existing ecosystems—for nature's sake and for our own.

I'd like to take this opportunity to acknowledge some people who have helped make this book a reality:

Sharron Dorr, Brenda Rosen, and Dawna Page at Quest Books—for imagining this project, inviting me to accept its challenge, and shepherding the publication process so smoothly and competently.

Dorion Sagan and Lynn Margulis—for their foreword, editorial suggestions, and corrections of a few factual errors in the manuscript.

My students and colleagues at New College in Santa Rosa (especially Deborah Kraft)—for listening as I explored some of the ideas for the book in lectures and for calling my attention to relevant Internet newsgroups and newspaper articles.

Readers of my monthly *MuseLetter,* with whom I shared several draft chapters—for offering encouragement and critical feedback.

Sandra and Peter Jensen and Trisha Feuerstein—for sharing their collections of books on biotechnology, as well as their critiques of my early draft chapters.

And finally my wife, Janet Barocco—for her patience and

encouragement during this past year. It's not easy living with someone obsessed.

—Santa Rosa,
Spring Equinox, 1999

Introduction

SUPPOSE YOU COULD redesign the genetic makeup of any plant or animal species on Earth, including *Homo sapiens,* to suit your whims. Would you?

For people who lived prior to the twentieth century, this question would have made no sense. No one knew what genes were or what they did. To be sure, plant and animal breeders had been carefully altering domesticated species for millennia, but hardly anyone imagined that it would someday be possible to reach into the very cells of a cow, a pig, a maize plant, or a human being and make minute alterations that would change the organism and its offspring in fundamental—and predictable—ways.

Today the question not only makes sense, it forces itself upon us. Indeed, the genetic redesign of life on Earth is already well under way. By mixing and matching genes, agricultural scientists are already engineering new crops that grow quickly, resist pests, and store well. And animal geneticists are hard at work producing "super cows" that yield more meat or milk, or that produce useful biochemicals in their udders.

Within decades—or perhaps only years—it may be possible

for wealthy parents to create "designer children" with specific characteristics: high intelligence, artistic talent, good looks, or immunity to certain diseases and addictions. Medical biotechnologists will grow human skin, blood, and replacement organs genetically matched to their intended recipients.

The biotech revolution inspires both intense hope and great fear. The hope is that, through gene-manipulating technologies, we will eliminate the "mistakes" of nature—diseases and limitations—not only in humans, but also in food crops and domestic animals. In the short term, we will banish hunger and want; eventually we will learn to clone the rare Einsteins and Mozarts who appear among us perhaps only once in a generation or once in a millennium and, by doing so, genetically enrich and uplift the entire human race.

The fear is that we will fail in our attempts and trigger a universal biological catastrophe; or that, even if we succeed, in doing so we will erode and destroy the human soul, and perhaps the very soul of nature.

Either way, the implications are enormous.

Every new technology—from the plow to the nuclear reactor—has posed unprecedented moral problems. Computers, for example, require us to think in unaccustomed ways about privacy and the social control of information. Prior to the invention of the computer, no one had to decide whether it was ethical for companies to compile and sell vast amounts of personal data (credit ratings, purchasing habits, and political views) about ordinary citizens. The invention of nuclear reactors and petrochemicals has forced upon us both practical and ethical quandaries having to do with toxic waste disposal: should toxic sites be located in poor neighborhoods—where workers live—or nearer to where the managers live? Who should pay the health costs entailed by increased cancer rates—investors or the community at large?

Biotechnology raises the moral stakes surrounding technological change to an entirely new plateau, but one with a kind of ancient, mythic resonance. By cloning and gene splicing, we are teasing apart and reshaping the very essence of life. And from the earliest times, human beings have believed that the acts of learning nature's secrets and of manipulating natural processes imply some kind of danger.

The creation story in Genesis tells of the first man and woman, expelled from paradise for eating from the Tree of Knowledge. Other familiar myths pursue similar themes: Icarus, whose father fashioned wings for him, flies too high and is dashed to the ground; Prometheus, who steals fire from the gods on behalf of humanity, is eternally punished by Zeus; Pandora, whose curiosity leads her to open a forbidden box, unleashes plagues upon the world. An appropriate "myth" of the modern age brings the message nearly up to date: in her 1818 novel *Frankenstein*, Mary Shelley tells of a well-meaning but arrogant scientist who, in his attempt to master the forces of life, dies at the hands of the pitiful monster he has made.

Clearly, human beings both hunger for power over the

natural world and at the same time distrust their own pursuit of that power. Recent advances in biotechnology are forcing us to confront this distrust as never before and to decide whether it is based in primitive superstition or primordial wisdom.

To Clone or Not to Clone?

The mythic resonance of the biotech debate is only deepened by the fact that events have brought it into sharp focus as we approach the end of one millennium and the beginning of the next.

In February 1997, the announcement of the birth of a sheep named Dolly—the world's first cloned mammal—provoked a firestorm of controversy. Could the same technique be used to make genetic copies of humans? Would the coming millennium see the realization of Aldous Huxley's *Brave New World*—a fictional dystopia in which batches of identical human embryos are grown in state-run "hatcheries"?

In December 1997, an eccentric physicist-turned-gene doctor named Richard Seed announced on national radio that he intended to open a human cloning clinic in the Chicago area. "Clones are going to be fun," he later blithely commented to an audience at a Chicago law school symposium on reproduction. "I can't wait to make two or three of my own self."

The prospect of human cloning immediately raised novel fears and hopes. Perhaps, said advocates, in the future people will be able to clone genetically compatible "spare parts"—hearts, livers, kidneys—in the laboratory, thus reducing the dangers and difficulties of organ transplants. Maybe infertile couples will have a new reproductive option. But critics of the idea wasted no time in summoning horrifying visions of a future inhabited by armies of identical, soulless, gene-perfect techno-humans.

Dolly's "creator," Scottish embryologist Ian Wilmut, declared that he was opposed to the idea of cloning humans, and several European nations rushed to ban any human application of the procedure.

The idea itself provoked so much public outcry that researchers, in response, began warning that an anti-cloning fever might set back the course of scientific inquiry, delaying all manner of important discoveries that could benefit humanity. Members of the International Academy of Humanists came down firmly in favor of continued unhindered research. Their joint statement—signed by Francis Crick (Nobel laureate and co-discoverer of the chemical structure of DNA), Richard Dawkins (Oxford University professor of the public understanding of science), Herbert Hauptman (Nobel laureate in chemistry), William V. Quine (Harvard philosophy professor), Simone Veil (former President of the European Parliament), and Edward O. Wilson (Harvard biologist and the father of sociobiology)—was published in the Summer 1997 issue of *Free Inquiry*. The authors concluded that "The potential benefits of cloning may be so immense that it would be a tragedy if ancient theological scruples should lead to a Luddite rejection of cloning."

On the other side, an unlikely anti-cloning coalition soon emerged, comprised of political and social conservatives (including prominent evangelical Christian Pat Robertson and right-wing columnist Charles Krauthammer), medical ethicists (including Ezekiel K. Emanuel), and radical ecologists (Jeremy Rifkin and Jerry Mander). Emanuel, a member of a presidential ethics commission charged with making recommendations on cloning, helped draft that group's recommendation—that federal laws be passed to prohibit human cloning, but that the laws be reviewed in three to five years.

Scientists predicted that the cloning of humans would be many years away. After all, the difficulties in cloning Dolly had been considerable. But only months later, in February 1998, the same team that produced Dolly reported that it had also cloned a calf. In May of the same year, Steven L. Stice, chief scientist at Advanced Dell Technology in Worcester, Massachusetts, announced that he and his colleagues were working to produce a herd of cattle genetically modified to be immune to mad cow disease. In July, a Japanese team claimed to have cloned twin female calves. Also in July, researchers in Honolulu stated that they had successfully cloned three generations of mice. By this time, the technique had been refined to such an extent that far fewer trials were required in order to yield a successful clone. The cloning of adult mammals was becoming more controlled and efficient—if not routine.

This fact was forcibly underlined in August 1998 when scientists at Texas A&M publicized their $2.3 million effort to clone the beloved pet dog of a multimillionaire couple. By now researchers were nearly unanimous in saying that the cloning of humans would be just a matter of time.

In September 1998, Richard Seed reported that he was actively working toward cloning himself and that his post-menopausal wife had agreed to be implanted with the cloned embryo and to carry it to term. The following month, researchers succeeded in transferring the genetic material from the egg of one woman to that of another—a feat that tested nearly all the techniques that would be required for human cloning, while stopping short of the act itself.

By the time you read this, it is possible that the first human cloning will already have been accomplished. The discussion will have shifted from "if . . ." and "when . . ." to "how shall we control this new ability?"

The Defining Moral Issue of Our Time

Though cloning has set off the most recent furor about applied genetics, a deeper and broader debate has been festering since the first successful gene-splicing experiments of the early 1970s. When Herbert Boyer (a researcher at the University of California Health Center in San Francisco) succeeded in transferring DNA from a toad into a living bacteria cell, alarm bells began sounding. For decades, scientists had enjoyed a kind of priestly status in American society and their discoveries were almost universally regarded as beneficial. Suddenly, biologists were speculating about the possibility of creating entirely new life forms impossible to produce through ordinary breeding—including plants with animal genes and animals with human genes. In 1976 Columbia University biochemist Erwin Chargaff, writing in the prestigious journal *Science,* asked: "Have we the right to counteract, irreversibly, the evolutionary wisdom of millions of years, in order to satisfy the ambition and curiosity of a few scientists?"[1]

Theologians decried the folly of geneticists "playing God"; science fiction writers spun tales of clones and bioengineered chimeras running amok; and ecologists warned of the possibility of the accidental release of a "designer" pathogen to which no plant or animal would be immune. The public outcry ultimately resulted in the creation of the modern bioethics movement which, during the last quarter of the century, has in turn spawned several advocacy groups and think tanks, and an entire department at the University of Pennsylvania.[2]

With the dawning of the Reagan era, the influence of the biotech critics waned: gene splicing was becoming commonplace and no disaster had yet occurred; besides, genetic research was now an investment opportunity, and Wall Street simply wasn't interested in discussing the morality of techniques that held so much promise of profit. In 1978 human in vitro fertilization (which used some of the same embryo-manipulation procedures that would later lead to cloning) was first successfully accomplished; within years it was considered a standard reproductive option. In 1980 the U.S. Supreme Court ruled that bioengineered living organisms could be patented. Quietly, leading-edge genetics research found commercial applications for recombinant DNA techniques, and by the end of the decade thousands of scientists in dozens of labs around the world were working to redesign plants and animals for profit.

But the moral questions about biotech did not go away; each new technical advance or practical application seemed to evoke them anew. In the late 1980s, scientists at Monsanto developed a recombinant version of bovine growth hormone (rBGH). By injecting rBGH daily into their cows, dairy farmers were able to increase milk production by up to 30 percent per cow. Critics pointed to studies showing that the treated cows had shorter lives and a greater tendency to develop mastitis (which required the increased use of antibiotics, residues of which ended up in the milk), and produced milk with elevated levels of a hormone called immunoglobin growth factor-1 (IGF-1), which has been associated with increased cancer rates in laboratory animals and humans. Public-interest groups demanded that milk from rBGH-treated cows be labeled as such, but the U.S. Food and Drug Administration, signaling its solidarity with the biotech industry, actually proposed banning such labeling. Consumer advocates began asking: Are corporations and regulatory agencies shoving biotech down consumers' throats?

At the same time, advances in human genetic screening were beginning to raise the specter of a new kind of eugenics—not a state-imposed Nazi racialism, but a "benign," commercially driven expansion of "choice" and "protection." Scientists began identifying DNA markers for various genetic diseases (sickle-cell anemia, Tay-Sachs disease, cystic fibrosis), and speculation swirled about the existence of a "gay gene" or a "criminal gene." Ethicists wondered: Wouldn't employers eventually want to know the genetic predisposition of potential employees? Clearly, genetic screening would soon raise difficult new privacy and discrimination issues. And wouldn't societies eventually become even more sharply divided along economic class lines, as parents who could afford genetic screening and enhancement produced ever more intelligent, strong, and healthy offspring, while everyone else continued to rely on nature's genetic roulette?

In December 1984, biologists James McGrath and David Solter had stated in *Science* magazine that "the cloning of mammals, by simple nuclear transfer, is biologically impossible." Biotech proponents had insisted that critics, in their ethical machinations about the cloning of humans, were wallowing in meaningless sensationalism; it could never really happen. Then, in 1997, came the announcement of Dolly's birth, along with Richard Seed's bombastic proposal to clone humans. It was clear that the critics were largely right. While their worst fears had not yet materialized, everyone could see that with the new gene technologies almost anything was possible. Moral and spiritual objections simply could not be avoided; the controversy was destined to flare up again and again.

Initial surveys showed that about half the American public opposed all uses of genetic engineering, even for the curing of a serious disease. About 85 percent rejected it for any use other than as a disease cure.[3] The idea of cloning humans proved so disturbing to so many people that, by late 1997, President Clinton had convened an ethics panel to advise him on the subject; the FDA had ruled that any effort in that direction would require agency approval; and nineteen European nations had signed a treaty saying that cloning people violated human dignity and was a misuse of science.

However reassuring these developments were for some biotech critics, it was clear that the pace of advances in cloning and other genetic techniques had at best been slowed somewhat. Humans would soon be cloned, regardless of national laws. Cloning clinics catering to the rich could be opened offshore; biotech insiders estimated that such clinics could be equipped for as little as half a million dollars each. And genetically engineered foods—unlabeled and untested—were quietly finding their way onto supermarket shelves.

What had yet to be put in place was a systematic way of looking at the moral problems of biotech. How could fundamentally new ethical questions be approached democratically? Where should society draw the line? And whose morality should be used as a yardstick?

A Spiritual Perspective

Throughout recorded history, in every known culture, people's ideas about right and wrong have been tied to religion—to sacred texts or the revealed will of supernatural beings. But during the past century or so, religion has taken a beating. Scholars have compared the beliefs and practices of various cultures and deconstructed ancient texts, revealing universal motifs and undermining the idea that any particular religion embodies some unique and absolute truth. Astronomers, physicists, and biologists have searched for evidence of God in telescopes and microscopes and come up empty-handed: the natural world apparently operates according to laws that are uniform and humanly comprehensible. And historians have reminded us of the dark side of religious history—the killing of millions for the sake of this or that "one true faith."

In response to these developments, most modern societies have mutated in a secular direction. Whether oriented toward a market economy or a socialist system of wealth distribution, all but a few have tended to reject religion as their basis for judging human behavior. And now at the end of the millennium, with socialism in retreat and the free market reigning triumphant, most people find it difficult to avoid the perception that everything is for sale and *nothing* is sacred. Efforts to forge a secular, humanistic system of ethics to replace traditional religion-based morality have met with limited success.[4]

However, despite the apparent triumph of secularism, polls continue to show that an overwhelming majority of people still believe in God. While they may define or characterize God in different ways depending on their religious backgrounds or affiliations, most poll respondents report belief in a reality beyond ordinary human experience, a reality characterized both by overwhelming power and inherent goodness, a source of love, truth, and beauty.

The public's religious expressions appear to be channeled increasingly along one of two routes: *religious traditionalism*, and what many regard as a *new spirituality*. These two approaches in turn yield a wide range of ethical standards relevant to the biotech debates.

By religious traditionalism, I refer to mainstream, orthodox, or fundamentalist movements associated with the great world religions—Christianity, Islam, Judaism, Hinduism, and Buddhism—and also with the indigenous spiritual traditions of native peoples around the globe. As we will see in chapter 5, traditional religious leaders are not united in their ethical stance toward biotech, and of course, not everyone within any given religion or movement agrees with its leaders.

Nevertheless, where objections are made they tend to be quite similar across religious and denominational lines. While Jews, Muslims, and vegetarian Buddhists may object narrowly that the genetic engineering of foods violates their traditional dietary rules, they and others—including fundamentalist Christians and Native American spiritual leaders—also tend to believe that biotechnology violates the sanctity of God's creation. Prince Charles of Britain speaks for this traditional, religious-based critique of biotechnology when he writes, "I happen to believe that this kind of genetic modification takes mankind into realms that belong to God, and to God alone. Apart from certain highly beneficial and specific medical applications, do we have the right to experiment with, and commercialise, the building blocks of life? We live in an age of rights—it seems to me that it is time our Creator had some rights, too."[5]

However, there are two reasons why traditionalism by itself offers limited potential for solving the moral dilemmas of biotech. First, secular advocates of the new technologies will inevitably say that religion-based moral objections are sectarian or parochial and apply only to the members of a given faith. Second, ethical judgments about biotech must be inferred from religious teachings that were formulated long before the advent of modern science. Moses did not bring back a commandment saying, "Thou shalt not clone." Jesus did not warn, "Woe unto you, ye generation of genetic engineers!" Mohammed made no specific mention of biotechnology, nor did the Buddha or Lao Tzu.

It is possible, therefore, that the new spirituality may provide a more promising avenue for moral evaluation of biotech than traditional religions. By new spirituality, I refer to a transdenominational movement stemming from three main sources. First is a growing public awareness of the history of religion (traceable in turn to recent New Testament scholarship and a widespread interest among Americans and Europeans—flourishing especially since the late 1960s—in Buddhism, Hinduism, and the shamanic religions of indigenous peoples such as the Native Americans). A second source arises from Jungian-related forms of psychology that regard the realm of "the sacred" not as a figment of human imagination but as an authentic and primary category of human experience. The last is the so-called New Age movement, which itself embraces a wide spectrum of interests ranging from natural healing to the channeling of messages from "ascended masters" and space aliens. Proponents of spirituality say that they are seeking to avoid religious dogmatism while at the same time honoring the innate human longing for meaning and for connection with some great, overarching pattern or force that transcends the purely material aspects of existence.

The new spirituality is evidently popular, as it is now the basis for an immensely successful literary genre typified by several of the best-selling books of the 1990s. However, while it has many leaders, none speaks authoritatively for the movement as a whole. This is because the new spirituality is not a monolithic—or even an easily definable—entity. If the spirituality movement could be said to have a core of universally agreed-upon tenets, that core might be the "perennial philosophy"—a phrase coined by Leibniz and used by Aldous Huxley as the title for a book published in 1944.[6] The perennial philosophy, according to Huxley, centers on the realization that there is more to us than just our physical bodies (with their genetic predispositions) and our environmental conditioning, and that life has meaning. According to this philosophy, every human being is a particularized expression of a universal sacred reality that we each strive to embody and express—a reality whose qualities are identifiable as compassion, justice, truth, and love.

The past quarter-century, during which the spirituality movement has burgeoned, has also been a time of widespread concern about environmental problems. And so the movement has mutated accordingly. Today most people sympathetic to the new spirituality would agree with Native Americans' long-standing beliefs that all of nature is sacred, that it is foolish arrogance to think that we can own the earth, and that we are each and all responsible for maintaining the integrity of the web of life.

These attitudes are directly relevant to the genetic engineering debate and strike a cautionary note. The human body is not a commodity, nor a collection of spare parts; all organisms belong to the integrity of nature, and we have no right to redesign them for our own profit or convenience.

However, as is the case with traditional religion, spirituality sometimes offers a mixed or uncertain moral assessment of biotechnology. Much of New Age philosophy is about self-improvement and about taking charge of our own spiritual growth. Many writers on spirituality say that humankind is evolving toward godhood and that the development of new technologies is merely the outer reflection of this inner process. How better to accelerate our maturation as a species than to take charge of the biological process of evolution? The spiritual-evolution cosmological model (traceable to Catholic theologian and palentologist Teilhard de Chardin, often described as one of the godfathers of the New Age movement) maintains that all of nature is seeking to perfect itself, and that humans have a special godlike role to play in this process. Could it be that we are destined to reshape nature, to perfect it, and to perfect ourselves in the process?

Suppose we could somehow produce a race of enlightened beings. Suppose we could isolate the genes that make people creative, intelligent, and compassionate. Suppose we could clone the Buddha. Wouldn't it be our spiritual duty to do so? Wouldn't our refusal be an act of evolutionary cowardice and failure?

The reverential and Promethean brands of spiritual thinking (as I will call them) could not be more different in basis and outcome. One emphasizes respect for creation as it is; the other promotes a vision of the purposeful transformation of ourselves and our world.

As we will see in chapter 5, both have roots that reach far back in time. Today, the ethical dilemmas posed by biotechnology force us to decide to what extent Promethean spirituality is an authentic revelation about our role in the cosmic scheme, or the result of confusing greed-based technological proliferation with a real evolutionary process; and whether reverential spirituality offers a reminder of the human need for humility before higher powers, or a superstitious brake on the attainment of our divine destiny.

And so, inevitably, the moral discussion that biotechnological developments force upon us is not only about biotech, but also about the meaning and future of spirituality.

Confronting Three Assumptions

Before we can enter into that discussion, it is important to name and confront three entrenched beliefs that would discourage us from engaging in it as wholeheartedly as we otherwise might. These beliefs are: (1) that it is already too late to stop or change the direction of the new technology; (2) that we should leave all such decisions to the experts; and (3) that all technologies are morally neutral, so we should only be talking about the *uses* of the new techniques, not the worthiness of the techniques themselves.

IS IT ALREADY TOO LATE? To date, hundreds of thousands of genetically engineered organisms have been created in laboratories around the world. Most people see themselves as mere bystanders in this process, reading occasional newspaper headlines about the latest technical advances. The genie is out of the bottle, say biotech promoters; it is impossible now—for economic as well as political reasons—to squelch these new industries, even if we wanted to.

However, biotech critics argue that it's never entirely too late for a society to deliberately change its relationship to a new technology; it just gets harder as time goes on. With regard to cloning and other procedures, we are still in the early stages of research and implementation. Thus it is important to make our ethical choices now. As we will see in chapter 7, activists and citizen groups are already having a noticeable effect in creating public policy regarding biotech.

SHOULD WE LEAVE THE DECISIONS TO EXPERTS? While many people agree that a moral discussion about cloning and biotechnology should be taking place, so far that discussion is occurring mostly among research scientists, leaders of the biotech industry, professional ethicists, and government regulators. There seems to be a good reason for limiting the debate to only these participants: even if it were possible to educate the masses about the subtleties of genetic science, there are few practical ways to include their opinions in the making of public policy.

Yet everyone in future generations will be affected by the decisions made today about genetic engineering—perhaps profoundly so. Thus, despite practical difficulties, it is essential that as many people as possible be included in the debate. In order to take part, people need to know the current state of affairs in biotech research and its likely development in the future. Perhaps they should also be acquainted with the resources of wisdom in the spiritual traditions of the world. If we are in the position of making a profound moral choice, it stands to reason that we should be asking ourselves: What would the Buddha, Jesus, Lao Tzu, or Mohammed have to say if confronted with these same choices?

While the ethical issues are new and complicated, we owe it to ourselves to help shape public policy from a perspective that is wise, humane, generous, and compassionate. We all must help decide, but we can do so intelligently only by acquainting ourselves with the arguments and evidence and by getting in touch with our deepest values.

ARE ALL TECHNOLOGIES MORALLY NEUTRAL? It is often argued that all technologies are morally inert and that only our use of them is constructive or destructive, right or wrong. It's easy to cite examples to bolster this point of view: the pistol that is used to murder a convenience store clerk might instead have been used to defend a home from burglars. Gun advocates would argue that, in the former case, it is wrong to blame the weapon itself, which is merely a lifeless tool; the problem lies with its user.

However, as historians of technology such as Marshall McLuhan and Lewis Mumford have pointed out, societies tend to mold themselves around their tools. The tool changes the user, often in ways that are unanticipated. For example, no one who has studied the history of America during the twentieth century doubts that the automobile has transformed our landscape and the relationship between cities and the surrounding countryside. Suburbs, malls, and interstate highways all owe their existence to the automobile. The national psyche has also been affected (imagine modern family life without the car), as has the economy (try picturing the twentieth century without Henry Ford's assembly line). Today we find ourselves grappling with seemingly unsolvable environmental, demographic, and economic problems resulting from our love affair with the automobile—problems that might have been lessened or averted altogether if we had taken the time earlier to think about consequences and alternatives.

Concerning the example cited above—that of guns—many people would argue that while moral responsibility rests with the gun user in any specific instance, the fact that more guns are manufactured and circulated within a society almost inevitably means that more people will be killed by them. Therefore, many people would argue, the society as a whole has a moral obligation to limit production of, or access to, such weapons.

The persistent use of a given tool reinforces in its user a certain way of thinking and of relating to the world. To a person whose only tool is a hammer, all problems look like nails. Thus it is important that we choose our tools carefully and discuss their potential side effects *before* altering our way of life so as to turn them into necessities of existence (as we have already done with the automobile).

Again, with regard to biotech, the time to ask tough ethical questions is *now.*

In this book I seek to explore the issues of biotechnology from a spiritual perspective that is as universal and inclusive as possible. This book is about biotech, but it's about religious tradition and spirituality, too. In coming to terms with this thorniest of moral dilemmas, I believe we can profoundly clarify understanding of our own inner nature, our relationship with tools and technologies, and our responsibility to the natural world.

Looking at biotech from a philosophical, moral, and spiritual perspective means trying to understand, not just the specific applications of biotechnology that are currently being deployed, but the assumptions and worldviews that those applications embody. What train of thought has led us to cloning and gene splicing?

CHAPTER ONE

The Spiral Ladder

IN ORDER TO understand the implications of biotechnology, we have to start with the fundamental facts and theories that underlie the new gene-manipulation techniques. At the very least, we need to define genes, chromosomes, and DNA and their respective functions.

However, in order to make a *spiritually informed assessment* of biotech, we need to look beyond textbook presentations of genetics. Molecular biology, in its development as a science, has largely sought to explain life processes solely in terms of chemistry and molecular activity—to *reduce* the whole to the interaction of its parts. This reductionist attitude of many molecular biologists is as useful as it is understandable, given certain historical and cultural factors that we'll discuss later both in this chapter and in the next. Nevertheless, reductionism—the idea that complex phenomena can best be understood by studying their simpler components—leads us away from spirituality, which views the world in terms of wholes and in terms of qualities, purposes, values, and ideals that cannot usefully be reduced to anything else.

As we will see, there are scientific alternatives to reductionism that offer a more holistic, integrative—even spiritual—way of looking at living things.

A Code of Life

Let's start with the science of genetics as conventionally understood and taught.

Within the nucleus of all living cells—except for bacteria, which do not have nuclei—are rod-shaped structures called chromosomes. The number of chromosomes in each cell varies with the species: human cells (other than sperm and egg cells) each contain twenty-three pairs of chromosomes; a fruit fly, in contrast, has only four pairs. For decades it was known that chromosomes were somehow involved in the conveyance of physical characteristics through heredity, but the exact nature of their role remained obscure.

In 1944, Oswald Avery and his colleagues determined that each chromosome is formed from a single convoluted cord of deoxyribonucleic acid (DNA) so densely packaged that it can be up to 100,000 times longer than the chromosome itself. Less than a decade later, James Watson and Francis Crick proposed a three-dimensional model for the DNA molecule, describing it as two interlocked spirals of submolecules.

The two complementary strands of the DNA double helix are related in much the same way as a photographic print and the negative from which it was made: each is capable of recreating the other. The molecule as a whole is composed of four kinds of molecular subunits called nucleotides, each containing a sugar, a phosphate, and one of four nitrogen-containing bases: adenine (A), guanine (G), cytosine (C), and thymine (T). The sugars and phosphates, linked end to end by strong chemical bonds, form the spiraling rails of the DNA staircase; the bases, projecting inward, are the stair steps of its central axis. Genetic information is encoded in the sequence of the base pairs making up the molecular staircase using an alphabet of only four letters (A, G, C, and T), in which adenine is always paired with thymine and guanine with cytosine.

The DNA code is essentially a set of instructions for the synthesis of proteins, which in turn do just about everything that takes place within the organism. In protein synthesis, the code embedded in a segment of DNA is copied onto a molecule of a similar substance called *messenger RNA* (ribonucleic acid). Messenger RNA is then "read" by other molecules called *transfer RNA*. These in turn bind to amino acids and assemble them in the order necessary to make a specific protein, which may then serve as an enzyme (which catalyzes cellular metabolic processes), as a structural material within cells, or as a molecular regulator for other DNA segments.

The DNA molecule is nature's ultimate chemical solution to the problem of information storage. It is so efficient that all the information needed to specify an organism as complex as a human being (three billion DNA "letters") weighs but a few billionths of a gram. While the DNA molecule is only about one millionth the diameter of a piece of string, all the DNA in a single human cell, if unraveled, would be more than six feet in length. All the information in every book ever written, if translated into

DNA, could fit easily in a teaspoon.

The data coded on a strand of DNA is localized in *genes*—a word coined in 1909 to refer to then-hypothetical units of inheritance. Today the term is understood to mean a segment of DNA encoding sufficient genetic information to assemble amino acids into a simple chain, or *polypeptide*. Polypeptides can in turn be linked to form more complex multichained molecules of protein. The length of genes varies; a typical gene might encompass one hundred turns of the DNA helix and include one thousand base-pair steps in the DNA staircase.

What do genes do? Plenty. Not only do they determine whether an organism will be an oak tree, a mosquito, or a humpback whale, they also specify individual physical characteristics, as well as size and rate of growth. Genes determine whether a particular person will have red or blond hair, brown or blue eyes, and so on; they may also destine a person to suffer an inheritable disease, such as Tay-Sachs disease or sickle-cell anemia. Recently, geneticists have even linked specific genes to aspects of human behavior, including tendencies toward anxiety, thrill seeking, shyness, or depression.

According to modern evolutionary theory, genes are also a key to understanding the emergence and diversification of life on Earth. Self-replicating molecules are believed to have first appeared in Earth's primeval seas nearly four billion years ago. Gradually, these first precursors of life drew together the chemical components necessary to form cells. Random mutations of this primordial genetic material, winnowed by natural selection, led to the evolution of more complex cells, then multicelled organisms.

Most mutations are harmful to the organism in which they occur, or at best inconsequential. But occasionally—whether through faulty copying of the DNA code during cell division, or injury to the chromosome from a stray cosmic ray or other environmental insult—an altered gene conveys some characteristic that enables the organism to survive better within its environment and leave more offspring than other members of its species. Rarely, this novel characteristic will lead to the genesis of an entirely new, stable species. Over hundreds of millions of years (according to theory), this process has resulted in the diversity of life as we know it.

In short, you are your DNA, and life is DNA's way of creating more of itself.

An Intelligent Context

It all seems so straightforward: genes record the computer program that shapes the development of all living things; just decode DNA and we'll know nearly everything there is to know about life, from its origin to its present workings.

However, many biologists say that the situation is not so simple. They point out that there is a tremendous amount we still don't know or understand about living things, from the molecular level upward. Where reductionist geneticists attempt to explain all the features of complex organisms and ecosystems in terms of molecules (DNA), the scientific critics point out that DNA is also conditioned by the organism itself and by the environment. Where most reductionist biologists treat living things as machines made up of interchangeable parts, their critics insist on a holistic, organism-centered or ecosystem-centered attitude.

One of the most prominent of these critics is Lynn Margulis, professor of biology at the University of Massachusetts, Amherst, and coauthor with Dorion Sagan of *What Is Life?* and other books. One of Margulis's claims to fame is her coauthorship, with British biologist James Lovelock, of the Gaia Hypothesis, which regards the entire planet as alive, in that aspects of the surface (atmospheric composition, temperature) are self-regulated by the metabolism and growth of its component populations. Margulis argues that the mechanistic or reductionist view of life is "misdirected, incorrect, or, at best, grossly inadequate."[1] Her own tendency toward "dynamic, interactive physiological thinking,"[2] which is shared by other critics of reductionist biology, might well be termed the Gaian Alternative.

Gaian thinking is systemic and ecological. When applied to genetics, it emphasizes the relationships, networks, and patterns of which genes are only a part and regards DNA as subservient to organisms and ecosystems as wholes.

Gaian scientists point out that what we know about life so far is extremely partial and limited. Sometimes our lack of understanding shows up in the form of discoveries that are surprising or difficult to account for. For example, once biologists had begun to discover the structure and functions of DNA, they were puzzled to find that about one percent of the cell's total DNA is *not* packed within chromosomes but is attached to the cell's surface. Some researchers have speculated that this out-of-place DNA comes from inside the cell or is carried by the bloodstream. Others think it may have something to do with the body's immunological responses. For now, its function remains a mystery.

At other times theories seem to lead in one direction and facts in another. For instance, assuming that DNA carries the codes governing the form and structure of organisms, we would naturally expect that cells with more genetic material would belong to more morphologically complex organisms, while cells with less genetic material would denote structurally simpler organisms. In some instances, this turns out to be true: bacteria, for instance, have far less DNA than hamsters. However, some plants have up to a hundred times more DNA per cell than humans, as do some fish and salamanders.

The apparent reason for this lack of correlation between an organism's size or physical organization and the amount of DNA contained in its cells is that many life forms carry around a lot of genetic material that has no known function. In humans, this "junk" DNA constitutes up to 97 percent of the entire genome. In mice, one apparently meaningless sequence is repeated a million times in each cell. But there is always the possibility that "junk" DNA actually has some purpose we simply haven't discovered yet.

Then there are occasions when our metaphors break down. The "computer program" metaphor for genes leads us to assume that each gene must encode one meaning and no other; that in differing species, similar genes encode similar meanings; and that similar structures in differing species must be specified by similar genes. But this turns out to be a gross oversimplification. Almost every gene studied in higher organisms has been found to affect more than a single trait. Many influence more than one organ system—a property known as *pleiotropy*. In the house mouse, for example, nearly every coat-color gene also has some effect on body size. To complicate matters even more, these pleiotropic effects often turn out to be species specific: in the domestic chicken, a single gene is involved in the development of some structures unique to birds—air sacs and downy feathers—but also of the lungs and kidneys, which are structures that occur in other vertebrate classes as well. This means either that the gene in question does something in chickens that is done by a different gene in other animals, or that the same gene does different things in different species.[3]

Richard Tapper, writing in *New Scientist* in 1989, underscored this ambiguous quality in the genetic code:

> Research at the forefront of molecular sciences has shown that we can no longer regard DNA—the stuff of genes—as a direct and complete set of instructions for the synthesis of proteins. The evidence begins to suggest that messages in DNA are, in themselves, no more precise than the symbols and sounds with which we communicate. As in the languages with which we are familiar, the correct sense of a message written in DNA seems to depend on the rigorous checking and correction of errors, and on the context in which they are read. . . . Thus genetic and evolutionary changes are no longer confined solely

> to the genome at the pinnacle of a hierarchy of information and control, but reside also in the interplay between DNA and the other components of cells.[4]

But how does an organism transmit "context" to its offspring? This, say the Gaian scientists, is a mystery we can only hope to solve by looking at relationships and wholes, not just at molecules in isolation. Furthermore, it is a mystery we should seek to clarify *before* we alter the biosphere by engineering DNA sequences in plants and animals on a commercial scale. Otherwise we could make mistakes with vast and permanent repercussions.

Curbing the Dogma

Much of the Gaians' critique of reductionist thinking centers on what is known as the "Central Dogma" of molecular biology, formulated by Francis Crick in 1958. According to the Central Dogma, genetic information flows from genes to proteins, never the other way around, according to this simple formula:

DNA RNA protein (enzyme) trait

Even in the 1940s there was evidence that the Central Dogma was overly simplistic and perhaps plain wrong. In her pioneering breeding experiments with maize, geneticist Barbara McClintock had discovered that genes regularly move between chromosomes. While the Dogma insists that genes are immune to their context,

McClintock (and subsequent researchers) have shown that "jumping genes" have different effects on the development of traits depending on their location in the chromosome. Though McClintock's work was largely ignored for nearly three decades (she received a belated Nobel Prize in 1983), the concept of "jumping genes"—or *transposons*—is now part of standard textbook knowledge.

Through the early 1970s the Central Dogma held sway in the thinking of most biologists. However, as findings have accumulated showing the importance not just of jumping genes but of overlapping genes, vanishing genes, silenced genes, entangled genes, and wandering genes, most research geneticists have abandoned the idea that discrete, identifiable entities called genes determine traits according to a simple, one-way formula. Instead, researchers now refer to "the fluid genome," which is as much an effect of the organism and its environment as it is a cause of the development of traits. Biologist Ernst Peter Fischer goes so far as to say that "There are . . . no genes! There are—at least in the cells of higher organisms—only pieces of genes, which *the cell can use when it makes proteins.* A gene is by no means a molecule that exists in the cell. Rather, a gene is a task that a cell has to accomplish."[5]

Philip J. Regal, internationally respected authority on plants and ecology, put it this way in a phone conversation with me: "The evidence [for the Central Dogma] is not very favorable. It might be better for bacteria than for other organisms, but certainly for multicellular organisms there are a lot of intervening steps between the DNA code and the actual organism itself. We have interactions between chains, we have what we call epistatic and pleiotropic effects, and then we have developmental processes that have their own emergent properties, so that the DNA code only influences those developmental processes, it doesn't direct them like puppets. There is no one-to-one reading of the genotype onto the phenotype. So they're way off base when they think that way and talk that way. It's good for getting grants, but it just is not in step with the rest of biology." [6]

Craig Holdrege, another Gaian scientist, expresses much the same point of view when he writes, "Regardless of how one defines genes, simple, monocausal, one-directional schemes as epitomized by the Central Dogma are no longer credible."[7]

Yet, strangely, *the Central Dogma refuses to die.* It still dominates the thinking of thousands of geneticists who are engaged in the creation of powerful new technologies for reengineering the natural world. Gaian scientists may point to mountains of research data as support for their systemic view of living things. But in commercial biotech labs and many universities, reductionist biology still reigns. Billions of dollars are invested on the assumption that snipping a gene here and inserting it there will lead to a predictable outcome. And popular books are written describing human beings—our personalities, foibles, and habits, no less than our body types and vulnerabilities to disease—as direct products of our genes.

Gaian geneticist Mae-Wan Ho of Britain's Open University writes: "The transition between the molecular genetic determinism of the central dogma and the new genetics is reminiscent of the transition between the separate, mechanical objects of the

Newtonian universe and the de-localised, mutually entangled entities of quantum reality."[8] For some reason, the scientists who are leading the biotech revolution have failed to make this important transition in their thinking.

The Rockefeller Foundation's Molecular Vision

The residual tendency toward reductionism on the part of bioengineers is not just intellectual intransigence. In fact, there are many reasons for the ascendancy of the mechanist philosophy. One of them came to my attention in my conversation with Philip Regal. I later confirmed what Regal told me by consulting *The Molecular Vision: Caltech, the Rockefeller Foundation, and the Rise of the New Biology,* an authoritative book by Lily Kay, historian of science at the Massachusetts Institute of Technology (MIT).

"This goes back to the 1930s, before there was any government funding for science," Regal explained. "There was then no National Science Foundation or National Institutes of Health. The big private foundations—especially the Rockefeller Foundation—funded everything. Around that time, the Rockefeller Foundation decided to rebuild biology on a new conceptual basis. They got a couple of physicists, Max Mason and Warren Weaver, to establish a new science—molecular biology." The term was coined in 1938 by Weaver, the director of the Rockefeller Foundation's natural science division.

"Two streams come together in this story," continued Regal. "One stream is the Rockefellers' infatuation with eugenics and the idea that society's problems are essentially genetically based, that the reason we have poverty and war and crime and dysfunctional families and alcoholism is bad genes. It was a kind of faith that came out of the larger eugenics movement. Their solution to agricultural problems was chemistry, and they saw the long-range solution of social problems coming from chemistry, too. All we've got to do is correct the genes.

"Of course, the Rockefellers had an economic incentive to put their faith in chemistry because they had made their money in pumping and distributing oil, the basis of the petrochemical industry. But they also had a political incentive to see human problems in chemical terms; this stemmed from their well-known hatred for the New Deal and Franklin Roosevelt. Their idea was to stop tampering with the political and economic system and wait for science to solve all the problems."

The eugenic philosophy was not original with the Rockefellers. According to Regal, "A lot of biologists were already thinking in eugenic terms anyway. Some of my older colleagues have told me, 'You know, we were all eugenicists back then.' It was a common way of thinking among biologists, chemists, and all sorts of other people. But the Rockefeller Foundation took it to an extreme. They were actually funding research in Germany, and later that of course became an embarrassment and the Rockefellers pulled out of their German projects in the nick of time when the racism became clearly insane.

"So that's one stream—the faith that everything is genetically determined. If you start out with that faith, then the science that you construct can have that built into it.

"The other stream had to do with Newtonian physics. The

people they hired—Max Mason and Warren Weaver—were physicists who really didn't know any biology. They were even more strongly committed to the deterministic, reductionistic ideology than the biologists were. The biologists were flirting with it, but Mason and Weaver were totally swept up in it. That has to do with the reason they left physics—which was because they were disgusted with quantum physics. The whole idea of uncertainty was totally against their grain. They had this old Newtonian idea of the billiard-ball universe: once we understand mechanics, then we can build upon that and everything will reduce to mechanics. Even Einstein had pointed out that that way of looking at things was a failure. After hundreds of years of hoping, electromagnetics did not reduce to mechanics."

Weaver and Mason were in fact leading physicists, but they had never accepted the new ideas of Heisenberg, Bohr, and the other quantum pioneers; they assumed instead that quantum physics would eventually be found to be basically flawed. Indeed, Weaver wrote in his autobiography that his faith in God led him to know as a moral fact that the world was very simple and that God would only build a world on the principle of certainty. And Weaver wrote of Mason that "his attitude [toward quantum physics] was more than mere avoidance or disregard: he actively disliked the subject, and considered that it was so unpleasantly messy, so full of internal contradiction, and so clearly headed in the wrong direction that he would have nothing to do with it."[9]

Regal continued: "So Weaver and Mason started bringing in people from physics and chemistry like Max Delbrück and Linus Pauling, who really didn't know any biology. They knew chemistry, basically. And they told these new recruits, 'We'll give you lots of money, and here is how you write your grant proposals.'

"They began a tradition of funding in which these people, and their graduate students, and their graduate students' graduate students, would write as though biology were a very simple sort of thing. The idea was, 'We'll crack the chemistry of the genes and this will lead to all sorts of benefits.'"

But the older school of biologists was not impressed. In fact, as Regal puts it, "There was a battle. Regular biologists said, 'Whoa, biology is really not that simple. Organisms are not that simple.' The molecular biologists laughed at them and said, 'We're going to show that biology *really is* that simple.' Then there were years of success. The molecular biologists found that the hereditary material was DNA, and it *was* a molecule. There was a universal code. And it seemed possible to understand the construction of proteins based on simple models."

Molecular biologists enjoyed decades of breakthrough upon breakthrough, looking at living things as though they were entirely explainable in terms of simple chemistry and ignoring the botanists, zoologists, mycologists, and ecologists who insisted that nature is complex. And so, in Regal's words, the molecular crowd "had closed their eyes and ears to what the rest of the biological community was saying a long time before genetic engineering ever came around."

One reason for the reductionists' success and apparent smugness lay in the fact that they had an inside track in fund raising. As Lily Kay extensively documents, all of the resources of the Rockefeller Foundation were put behind selling the molecular philosophy. She writes:

> This new science did not just evolve by natural selection of randomly distributed disciplinary variants, nor did it ascend solely through the compelling power of its ideas and leaders. Rather, the rise of the new biology was an expression of the systematic cooperative efforts of America's scientific establishment—scientists and their patrons—to direct the study of animate phenomena along selected paths toward a shared vision of science and society. . . .[10]
>
> The pre-World War II level of Rockefeller Foundation support for molecular biology amounted on average to about two percent of the entire federal budget for scientific research and development. This figure gains significance when we consider that the lion's share of government support for the life sciences went to agricultural research. When we also include the indirect effects of the Foundation's support for molecular biology in Europe and its massive support for biomedical research, the financial resources for molecular biology become even more impressive. It is clear that the Rockefeller Foundation was in a strong position to shape fields in life science. . . .[11]

The Rockefellers were at the same time advancing the idea among their grant recipients in the social sciences that molecular biology was going to solve a wide range of human problems. And Foundation scientists were routinely receiving Nobel prizes and National Academy of Sciences appointments.

"So, from the point of view of the molecular biologists," Regal told me, "the money was coming in, and they could talk university officials into hiring molecular biologists instead of traditional biologists. They were taking over biology departments right and left. They were being wined and dined by people in industry. And soon the military wanted to know what was going on. As Susan Wright recounts in her book *Molecular Politics,* the molecular biologists were getting a lot of encouragement early on because of government interest in germ warfare.

"They're flying first class, they're doing just fine. So why *should* they listen to the rest of the biological community? Actually, they see the rest of the biological community as an anachronism. The slogan I've heard over and over was, 'We've got to tear biology down and put it on a proper scientific basis,' which was chemistry and physics. That was the agenda—to tear down the old neighborhood and build this new high-rise. They assumed that the people who actually worked with real plants and animals and their habitats were old fashioned and philosophically unsophisticated. But because the molecular biologists refused to listen, they couldn't see how narrow their own outlook was becoming. They literally didn't know what they were missing, and there's been no reason for them to listen because they've been so successful."

During World War II, the Rockefeller Foundation was instrumental in helping the war effort. After the war, its directors persuaded the government to take over many of its programs, so that future financing would come from taxpayers. With the cold war in full freeze, Congress discussed the idea of promoting science at a much higher level and set up the National Science Foundation (NSF) and the National Institutes of Health (NIH). "Basically those took over the Rockefeller Foundation's programs," Regal explained, "and much of the philosophy that was behind them. So we've actually got this philosophy built into a lot of our federal programs.

"That makes it even more difficult to challenge the current direction of genetic research and technological implementation. Many of these bureaucrats are not really deep thinkers; they're process people. Once the system is in place their job is to keep it going, and they don't sit around and try to rethink its basic philosophy."

This Is Not Merely Academic

The story of the Rockefeller Foundation's influence on the early development of molecular biology throws a cold light on recent discoveries about the role of DNA molecules in shaping organisms.

Whether reading the pages of the *New York Times* science section or watching a television documentary, one seldom encounters the words of a critic of the molecular vision—of someone who might seek to put the role of DNA in perspective as merely a part of a more complex system involving the entire organism and its habitat.

Nor is one ever informed that the science of biology is largely divided into two camps: on one side, most ecologists, botanists, zoologists, and mycologists—many of whom might be described as Gaians—who work directly with live organisms, view nature as complex and self-directed, and have become systematically marginalized from the scientific debate; and on other side, molecular biologists—most of whom view nature as essentially simple and have taken over many university biology departments, monopolized most sources of funding, and garnered the lion's share of recent Nobel prizes in the life sciences.

Further, the casual reader of science articles is spared the confusing news that the basic science underlying genetic engineering is controversial *even among geneticists.* While many still adhere to the Central Dogma, others have come to acknowledge that the initial expectations of genetic determinism have not been confirmed, and that the evidence—far from confirming simple, predictable, one-way effects from genetic change—instead reveals a heterogeneous reality in which genes jump; in which DNA is continually being altered by the organism and its environment; and in which a particular DNA sequence may have a variety of functions that are turned on or off depending on its place in the genome and a host of other factors, many of which are still unknown.

News stories about the latest advance in genetic engineering rarely contain comments from renegade molecular biologists like Dr. John Fagan, who in 1994 returned $614,000 in grant money to the National Institutes of Health because he had come to believe that genetic engineering is inherently dangerous and unethical; or Richard Strohman of the University of California at Berkeley, who was named as a plaintiff in a lawsuit to force the

FDA to label genetically altered foods. Only readers who go out of their way to subscribe to on-line organic agriculture or pure-food activist newsgroups are likely to learn about the debates over ecological hazards from escaping genes, damage to pollinators from engineered crops, or unexpected allergic reactions in humans from transgenic foods.

If this were merely an academic dispute, one might conclude that the reductionists and genetic determinists have arrived at a partial, oversimplified view of the world which they promote with remarkable enthusiasm and effectiveness—and let it go at that. After all, science has seen its share of intellectual fads and fashions, so we should expect that, in the course of time, the pendulum of opinion will swing and biological reductionism will give way to Gaian holism.

However, the dispute is *not* merely academic. The molecular vision of life—as pursued by the genetic engineers—is beginning to affect the real world. A substantial portion of the foods we eat are already genetically engineered; the same is true of pharmaceuticals. Domestic livestock animals are actively being "improved" by the addition of DNA sequences, and application of similar procedures with humans is only months or (at most) a few years away. Humanity is engaged in the reengineering of a substantial portion of life on earth—on the basis of a set of assumptions that has already been tested and found wanting.

A highly partial world view has subtly come to dominate science and society. And today it even influences the way we view ourselves, our thoughts, and our behavior.

Genes and Behavior

If genetic determinism and reductionism give an incomplete account of the origin and development of physical traits in organisms, they are even more misleading when offered as explanations for human behavior and mental processes.

In recent years, studies of human twins and discoveries in molecular biology have shown that many behavioral traits may be linked to genes. Researchers have claimed that, in our own species, alienation, extroversion or introversion, traditionalism, leadership aptitude, religious conviction, tendencies toward optimism or pessimism, and vulnerability to stress all seem to be *at least partly* genetically programmed. If this is true of behavioral *tendencies*, couldn't behaviors themselves—even intricate ones—also be chemically scripted?

The links between genes and behavior are tantalizing. Might it eventually be possible to understand a given species' entire behavioral repertoire in terms of genes and to alter its behavior by cutting and pasting DNA segments? Could we also come to understand the genetic basis of violent and criminal behavior in humans and create a molecular cure for social ills?

During the past decade, this trend of thinking has become extremely popular in scientific academia and in the popular press and has led to the birth of two influential new disciplines—sociobiology and behavioral genetics. Dean Hamer, the originator of behavioral genetics, writes that "the latest research in genetics, molecular biology, and neuroscience shows that many core personality traits are inherited at birth, and that many of the differences between individual personality styles are the result of differences in genes."[12] E. O. Wilson, the father of sociobiology, stretches the point further by theorizing that "Human behavior—like the deepest capacities for emotional response which drive and guide it—is the circuitous technique by which human genetic material has been and will be kept intact."[13] The message is clear: genes are primary and determinant; human behavior, emotions, and mental processes are secondary and derivative.

As might be expected, Gaian biologists are skeptical about the claims of sociobiologists and behavioral geneticists. No one doubts the statistical links between the presence of certain genes and the tendencies toward specific behavioral traits. However, Gaian biologists insist that these links have been dramatically overstated.[14]

Are genes the ultimate cause, or just links in a complex chain? Is the influence one-way, or can genes rearrange themselves because an organism alters its behavior patterns? Does the environment exert an influence? The Gaians claim that the evidence portrays genes as aspects of an integrated system and that by isolating them and giving them central importance, sociobiologists and behavioral geneticists paint a highly misleading picture.

If genes were the sole, mechanical determinants of traits, as the Central Dogma implies, then behavioral genetics would have a strong case. But it appears that organisms possess reserves of information other than DNA, even at the cellular level. The more complex the organism, the more evident it is that there is something capable of transmitting information from one generation to the next that is not reducible to the letters of the genetic alphabet.

If we want to explain behavior by genes, a good place to start is with instincts. After all, it would seem far easier to account genetically for the apparently rigid, inborn, scripted behaviors of animals than for the fluid, adaptable, social behavior of humans.

Virtually all vertebrate animals (and many invertebrates) display instinctual behaviors, some of remarkable complexity. Scientists have long wondered how instincts are passed from generation to generation and, lacking any other explanation, have appealed to genes as a likely mechanism.

Molecular biologist Margaret Mellon of the Union of Concerned Scientists told me, "There's a lot of information that suggests that genes play some role in behavior. If you're a termite and you've had certain genes knocked out, then you won't be able to follow the termite in front of you. But when you get to mammals and other creatures with more complex behaviors, then there are increasingly rich interactions between the environment, the way animals are raised, and their genes. It's a complicated interaction."[15]

One problem with the idea that genes script instincts in a simple, understandable way is that in some cases it is difficult to conceive of any DNA sequence that could contain and transmit the observed behavior patterns. Consider the migration of the European cuckoo, which lays its eggs in the nests of other bird species. Its young, hatched and reared by surrogate parents, have no opportunity to learn from adult birds of their own kind. Near the end of the summer, adult cuckoos migrate to their winter habitat in Southern Africa. About a month later, the young cuckoos congregate and then fly together to the same region of Africa, where they join their elders.

Termites offer an equally impressive example of instinctive behavior. The nests of species in Africa and Australia can be up to ten feet high, with well-defined features that suggest the existence of something like an architectural blueprint in the "minds" of their builders. Hundreds of thousands of tiny, blind insects, which have no known means of detailed communication, coordinate their efforts to build effective channels for cooling and ventilation, and orient their towers to maximize exposure to morning and afternoon sun and minimize midday heating. It's easy to see how genes for chemical sensitivity could help termites follow one another along narrow paths, but it's not so easy to see how genes could code for all the planning and coordination necessary to build the insect equivalent of a high-rise office building.

These complex, scripted activities are qualitatively different from a mere behavioral predisposition—toward anger, left-handedness, or thrill seeking, for example. It is one thing to say that such instincts "must be" genetically programmed, but quite another to show the specific connections between DNA base pairs and a knowledge of migratory routes or nest architecture. So far, no animal behaviorist has recorded a case of genetic mutation giving rise to a new behavior. Will we ever be able to snip an appropriate DNA sequence from a cuckoo cell, insert it in a bat embryo, and produce a bat that migrates cuckoo-like? While the question remains open at present, it seems doubtful.

Gaian scientist Robert Wesson, emphasizing the unimaginable intricacy and systemic integration of the organism, argues against crude, one-way explanations of the origin of behaviors in genes: "To alter the running forelimb of an insectivore to the digging arm of a mole," he writes, "required a large number of changes proceeding concurrently in bones, muscles, tendons, feet, and so forth. Genes, having to operate as an integrated system or a team, have to change as a team, and no one has any idea how this is managed. . . . To develop elaborate behavioral patterns [i.e., instincts] is far more complex than sculpturing an organ. It requires not only making exceedingly delicately organized structures, the branching neurons, but also dictating connections among them and the organization of their operations, as though composing computer software." Wesson concludes with what may be an understatement: "We are nearly as far from understanding the origin of life's complexities as the origin of the universe."[16]

Nevertheless, reductionists—finding a statistical correlation between a gene and a behavioral tendency—immediately draw the most grandiose conclusions. In the end, they say, we humans are nothing but automatons controlled by our "selfish genes," whose only imperative is to reproduce and proliferate at the expense of other selfish genes. Even altruism is reducible to "inclusive fitness," or genetic selfishness (sociobiologist Michael Ghiselin writes that "'altruism' is a metaphysical delusion" and "what passes for cooperation turns out to be a mixture of opportunism and exploitation").[17] Gaian scientists sharply dissent, noting that efforts to explain away cooperation and altruism in terms of the supposed anthropomorphic motives of genes can succeed only by ignoring overwhelming evidence showing that, in evolution, cooperation is at least as fundamental as competition.

Mae-Wan Ho strikes a combative note in this regard when she writes that the attribution of social behavior in animals and humans to genes selected in evolution "has produced a veritable industry for many third-rate scientists with limited imaginations who can think of nothing better to do than dream up selective advantages for putative characters controlled by putative genes, thereby becoming an instant success with their professors as well as the darlings of the equally simple-minded science journalists writing for the popular media."[18] Robert Wesson concurs. "Sociobiology," he notes wryly, "tells us more about the sociology of biology than about the biology of society."[19]

Genes and Evolution

Reductionism and genetic determinism—again, despite their conflicts with new discoveries about the fluid nature of the genome—also dominate most academic discussions of how life emerged and how living things came to be the way they are.

With rare exceptions, only fundamentalist Christians question the reality of biological evolution in some form. But what form? The term *evolution* essentially signifies only "change proceeding over time." Darwin supplied a hypothetical process (random variation and natural selection), but knew nothing about genes and mutation. The French naturalist Jean-Baptiste Lamarck (1744-1829), who coined the term "biology," proposed a theory of evolution in which characteristics acquired by an organism—such as skills helpful in obtaining food, escaping predators, or attracting mates—can be passed on to offspring. Lamarck's ideas were criticized by Darwinists and were out of fashion for more than a century, but now there are neo-Lamarckian evolutionists (such as Robert Wesson) who argue that there is strong evidence that he was, at least in part, correct. Darwin argued that evolutionary change must be slow and gradual, but fossil evidence of sudden evolutionary jumps has led paleontologists Stephen Jay Gould and Niles Eldridge to propose the theory of "punctuated equilibrium." In short, rather than consensus on the facts of the evolutionary record, there are many theories of how that evolution occurred.

For most of this century, a gene-centered theory of evolution has dominated most discussions about the origin and development of living things. Proponents of this "neo-Darwinian synthesis" have regarded DNA as central to the evolutionary process. In their view, random mutations are the engine of evolution, while natural selection turns the steering wheel. For example, a plant species doesn't just decide to grow a certain type of colorful flower in order to attract pollinators. Rather, over time, it produces a wide range of random genetic changes, and the changes that happen to help the plant reproduce most successfully (which may include colorful flowers) tend to be preserved in succeeding generations.

In reality, the role of DNA in evolution is still largely mysterious—beginning with the molecule's origin. Both DNA and its theoretical evolutionary precursor, RNA, are extremely complex chemicals. How did they first appear in Earth's primeval seas? Despite decades of experiment and speculation, no one knows. Some (though not all) of the amino acids necessary for protein synthesis have been produced spontaneously in experiments designed to mimic Earth's presumed primordial environment. But no such experiment has yielded nucleic acids, fats, starches, or proteins—and these key materials together make up 90 percent of the dry weight of a bacterial cell. Moreover, one would naturally expect that the simplest form of self-replicating object would be the part of the DNA program that says only "copy me." This is essentially what a virus is. But a virus cannot replicate outside of a living host cell; therefore, the first virus must have arisen *after* the first cell, not before.

The notion that evolution proceeds on the basis of purely random genetic mutations winnowed out by natural selection seems simplistic on many counts—particularly so in the face of a ground-breaking series of experiments by British biologist Dr. John Cairns and colleagues at Harvard University in 1988, later replicated by Dr. Barry Hall of Rochester University, among others.

Cairns experimented with the species *E. coli,* which is normally incapable of metabolizing lactose. When exposed to the sugar, the bacteria soon gave rise to descendants able to metabolize it. This adaptive response was different from anything that could be attributed to purely *random* mutation and selection; Cairns concluded that a "directed mutation" appeared to be occurring, and that "populations of bacteria . . . have some way of producing (or selectively retaining) only the most appropriate mutations."[20] In other words, the tiny organisms were somehow able to come up with just the mutations they needed.

In a separate experiment, Barry Hall started with a strain of *E. coli* unable to synthesize the amino acid tryptophan. He gave them tryptophan for a few days, then deprived them of it. The number of bacteria that mutated to produce tryptophan on their own jumped as much as thirty times the normal mutation rate. Hall documented that the mutation did not occur until needed and did not occur in bacteria starved of other amino acids. He later commented, "I can document [directed mutations] any day, every day, in the laboratory."[21]

Cairn's and Hall's findings have provoked a firestorm of controversy among biologists. A. S. Moffat wrote in *American Scientist*:

> The stakes in this dispute are high, indeed. If directed mutations are real, the explanations of evolutionary biology that depend on random events must be thrown out. This would have broad implications. For example, directed mutation would shatter the belief that organisms are related to some ancestor if they share the same traits. Instead, they may simply share exposure to the same environmental cues. Also, different organisms may have different mutation rates based on their ability to respond to the environment. And the discipline of molecular taxonomy, where an organism's position on the evolutionary tree is fixed by comparing its genome to those of others, would need extreme revision."[22]

Laboratory evidence suggests that fruit flies also experience directed mutations, and the observed rapidity with which insects and rats acquire immunities to poisons implies that this phenomenon may be universal. The nature of the process whereby organisms direct their own mutations is still a mystery, but some aspects of it are becoming clearer. Clearly, the duplication of genes, the alteration of regulatory genes, and the increase of mutation rates under stress are involved. But even as the details are revealed, the fundamental mystery deepens: *What within the organism is capable of speeding up, slowing down, or "directing" mutations?*

Many mechanist biologists prefer to ignore directed mutation altogether since it violates the neo-Darwinian notion that genetic mutations must be random. Those who do acknowledge its reality insist that the genome alone is responsible for producing changes within itself. But this seems more an article of faith than a reasoned explanation.

Gaian biologists like Ho argue that the entire organism, in association with its environment (which includes many other organisms), is ultimately responsible for producing needed genetic change. The genome, rather than being the sole motivating force in the evolutionary process, is merely one aspect of a complex system with many levels of organization and multiple internal and external feedback loops. Reductionism teaches that the genome produces the organism which—through selection and across many generations—adapts to its environment. But the organism also produces the genome, and the environment shapes the organism. There is downward causality (from the environment down to the gene) as well as upward causality.

Physicist Fritjof Capra notes: "The Gaia theory . . . [has] exposed the fallacy of the narrow Darwinian concept of adaptation. Throughout the living world evolution cannot be limited to the adaptation of organisms to their environment, because the environment itself is shaped by a network of living systems capable of adaptation and creativity. So, which adapts to which? Each to the other—they *coevolve*."[23] The living system is thus capable of being influenced both from within and without. As Robert Wesson puts it, "Evolution is the result of at least four major factors—environment, selection, random-chaotic development, and inner direction—and one might no more expect any law to govern it all than to find a law of the mind."[24]

The Gaian Alternative

If the neo-Darwinian synthesis and genetic determinism provide an inadequate conceptual basis for understanding organisms, what do the Gaians offer in their place?

Mechanistic-reductionist biology is based on the belief that all properties and functions of living things will eventually be explained in terms of the properties of their nonliving constituents. There are at least two obvious alternatives to this belief. One is to say that living things contain some force, field, substance, or entity that is *not* present in nonliving matter. This approach to the problem is known as *vitalism*. The other alternative is to propose that the living organism as an integrated whole cannot be fully understood from a study of its parts, because the organizing relations within the organism produce novel, emergent effects that even a thorough knowledge of its chemical constituents cannot predict. This latter approach is known as *organismic* or *living systems* theory.

Vitalism achieved considerable influence in the early twentieth century through the work of German embryologist Hans Dreisch. When Dreisch destroyed one cell of a two-celled sea urchin embryo, the remaining cell developed, not into half a sea urchin, but into a complete smaller animal. Dreisch concluded that there was something about living organisms that remained whole even though its parts were removed or damaged. To explain this "something," Dreisch postulated a nonphysical causal factor he called *entelechy,* which organizes and controls physicochemical processes during reproduction, morphogenesis, and regeneration. He argued that the growth of an embryo could be *affected* by genetic change, but it could never be fully *explained* in terms of genes, proteins, and amino acids.

Entelechy is a Greek word meaning "that which contains its goal within itself." Dreisch believed that while genes can specify the chemical makeup of the organism, the organization of its cells, tissues, and organs is determined instead by entelechy, which is inherited nonmaterially from past members of the species. Neither a form of matter nor energy, entelechy is nevertheless somehow capable of acting upon the complex living systems under its control.

Dreisch's critics seized upon this "somehow" as the weak link in his chain of logic. How can something nonmaterial act upon material things? Moreover, they cited new evidence from molecular biology showing that complex structures like ribosomes and viruses assemble themselves much the way crystals do. While living organisms are far more complex than ribosomes or viruses, perhaps this is a difference of degree, not of kind; in that case, it would be pointless to hypothesize a nonmaterial causal entity. The appeal to such a nonphysical agent raised a philosophical problem that few scientists wanted to contemplate: if entelechy exists, its nonphysicality means that there is no way to demonstrate its existence, except by inference. If science were to start admitting the reality of such entities, would ghosts and spirits be next?[25]

Vitalism virtually disappeared from serious discussion for several decades, but in the 1980s British cell biologist Rupert Sheldrake advanced a new quasi-vitalistic theory of formative causation that appeals to *morphogenetic fields* as an explanation for the complexities and regularities of nature. Morphogenetic fields, in Sheldrake's view, are nonmaterial "probability structures" consisting of direct connections across space and time between similar entities and governing the form and behavior of everything from simple molecules to entire communities of complex living beings.

Nature, according to Sheldrake, tends to develop habits. "By morphic resonance, the forms of all similar past systems become present to a subsequent system of similar form. . . . The probability structure of a morphogenetic field determines the probable state of a given system under its influence. . . "[26] Whereas reductionist biology "attributes practically all the phenomena of heredity to the genetical inheritance embodied in the DNA, according to the hypothesis of formative causation organisms also inherit the morphogenetic fields of past organisms of the same species. This second type of inheritance takes place by morphic resonance and not through the genes."[27]

In contrast to previous vitalistic theories, formative causation generates testable predictions which are the subject of ongoing research by Sheldrake and others. Sheldrake asks, for example, whether termite colonies are organized by morphic fields. To find out, he suggests housing one part of a colony within a portable structure that can be taken away from the rest of the colony and then observing whether the termites' behavior remains coordinated despite the lack of any physical means of communication.[28]

Organicism offers a different solution altogether by abandoning the notion of a nonmaterial causal factor in living things, and appealing instead to the emergent properties of systemic organization as the source of organisms' unique powers of regeneration, self-repair, and reproduction. Two influential pioneers in this approach are Chilean neuroscientists Francisco Varela and Humberto Maturana, who began publishing their work in the early 1970s. Varela and Maturana coined the word *autopoiesis*—from the Greek words *auto* (self) and *poiesis* (to make)—to denote the dynamic, self-producing, and self-maintaining activities of organisms. In their view, the *organization* of a living system is a set of relations among its components; autopoiesis is a general pattern of organization common to all living systems, but every subset of living systems (e.g., all bacteria, all maple trees, all bumblebees, all rainforest ecosystems) has its own unique patterns of organization as well. The system's *structure,* in contrast, consists of the actual chemical components and their relations. Given that immediately after death the material constituents of any organism are the same—and that organisms at absolute zero temperature (0° Kelvin) may be entirely viable and reproductively competent—*life* can not reside in the material itself. Maturana and Varela emphasize that the system's organization is independent of its structure, so that a thorough description of the components of a living thing (including its DNA) could never predict all the properties that arise from its pattern of organization.

Autopoietic entities have boundaries and engage in self-maintenance and reproduction; they ingest raw materials and energy from their surroundings and use these to make and replace their own components. But despite their continual cycling of materials, autopoietic entities maintain a coherent organization. It is this organization—rather than the stuff of which they are composed—that is the entities' real source of identity.

Fritjof Capra, in his book *The Web of Life,* suggests that organicism and living-systems thinking may serve as the basis for a new synthesis uniting biology, physics, and ecological thinking. "A theory of living systems," he writes, "consistent with the philosophical framework of deep ecology, including an appropriate mathematical language [deriving from complexity and chaos theory] and implying a nonmechanistic, post-Cartesian understanding of life, is now emerging. . . . The new synthesis of mind, matter, and life . . . involves two conceptual unifications. The interdependence of pattern and structure allows us to integrate two approaches to the understanding of nature that have been separate and in competition throughout Western science and philosophy. The interdependence of process and structure allows us to heal the split between mind and matter that has haunted our modern era ever since Descartes."[29]

While organismic theory, like mechanistic biology, avoids postulating the existence of nonmaterial factors like entelechy, it nevertheless looks at wholes as being more than the sum of their parts. Organicists do not deny that DNA plays an important part in the growth and maintenance of living things, but insist that it is only a part. This contrasts with the reductionists' and neo-Darwinians' focus on the smallest identifiable component of the organism as the key to understanding the whole. "The driving force of evolution, according to the emerging new theory," writes Capra, "is to be found not in the chance events of random mutations, but in life's inherent tendency to create novelty, in the spontaneous emergence of increased complexity and order."[30]

While vitalism and organicism are clearly distinct ways of thinking, both have spiritual overtones. Entelechy can easily be compared—if not equated—with the idea of a spirit-force, some version of which is to be found in many cultures (Polynesian *mana;* Chinese *chi;* Japanese *ki;* Indic *prana;* to name a few). The concept of morphogenetic fields suggests the existence of a transcendent reality beyond the physical world. And the autopoietic organization of living things suggests an implicit capacity within nature for the development of mind. "Living systems are cognitive systems," writes Maturana, "and living as a process is a process of cognition";[31] hence, consciousness is not an accidentally conceived genetic survival strategy, but a fundamental characteristic of living matter and an emergent potential inherent within the universe from its beginning.

All of this is worlds away from the reductionist view of life and mind typified by comments like the following by sociobiologist Michael Ghiselin: "We have evolved a nervous system that acts in the interests of our gonads, and one attuned to the demands of reproductive competition."[32] The mind, for reductionists, is an adjunct of the body, whose purpose in turn is solely to perpetuate and replicate genes. Culture and religion are likewise to be understood as collective efforts to maximize the survival of genes.

Modern physics and organicism share a willingness to contemplate the possibility that mind represents more than accident or genetic strategy. Compare Ghiselin's statement with the following one by physicist Freeman Dyson:

> The mind, I believe, exists in some very real sense in the universe. But is it primary or an accidental consequence of something else? The prevailing view among biologists seems to be that the mind arose accidentally out of molecules of DNA or something. I find that very unlikely. It seems more reasonable to think that mind was a primary part of nature from the beginning and we are simply manifestations of it at the present stage of history. It's not so much that mind has a life of its own as that the mind is inherent in the way the universe is built, and life is nature's way to give mind opportunities it wouldn't otherwise have.[33]

The Gaian world-picture (particularly, perhaps, that of the organicists) is in harmony, not only with the new physics of quantum theory, complexity, and chaos theory, but also with the actual experience of field biologists. As Margulis notes, "The study of physiology and immersion, especially in tropical nature, tends to lead students to a perception that the living planetary surface behaves as a whole. . . ." Yet armed against this perception, says Margulis, reductionists mired in 19th-century thinking serve as "academy guards" who, "using neo-Darwinism as an inquisitory tool, superimpose a gigantic super-structure of mechanism and hierarchy that protects the throbbing biosphere from being directly sensed by these new scientists—people most in need of sensing it."[34]

Sadly, as noted earlier, the casual reader of newspaper or general-interest magazine articles on science may never notice any alternatives to prevailing reductionist biology. Discussions about evolution are nearly always cast in neo-Darwinian terms, and the latest findings in genetics or advances in cloning techniques are inevitably reported with a genetic-determinist slant. This stubborn habit of thought, present especially in English-speaking countries, that seeks relentlessly to explain away the wholeness and meaning of life, mind, culture, and spirituality was never grounded in good science. It was, and is, a knee-jerk reaction to what is perceived as traditional, dogmatic, antievolutionary religion that only pretends to objectivity.

CHAPTER TWO

Machines and Organisms: A Spiritual Perspective

GENETIC ENGINEERING. BIOTECHNOLOGY. These terms portray the marriage of two profoundly dissimilar worlds. On one hand is the world of life—the squirming, breathing, eating, chaotic, inner-directed world of plants, animals, fungi, and bacteria. It is the world of forests, tide pools, and prairies. It is also the world of human experience—of mind and imagination, of family and community, of spirituality and tradition, of hating and loving.

On the other hand is the world of machines—the hard, metallic or plastic, dead, mechanical world of designed and manufactured objects, of invention and patent rights, of investment and profit, of novelty and obsolescence, of commercial utility.

The words tell a story: as we adopt *genetic engineering* and *biotechnology,* our interactions with living things increasingly become dominated by a mode of thinking conditioned to the design, manufacture, selling, and use of machines. This transformation is taking place first in the minds of the mainstream geneticists, who (as we saw in chapter 1) have adopted a reductionist-mechanist attitude toward their object of study.

The reasons how and why this has happened—and why the

Gaian alternative has made so few inroads into the academic and commercial life sciences—are manifold. One significant reason is the well-known battle between Darwinian evolutionism and Bible-based creationism that began in the 19th century and is still being fought on radio call-in shows and in school-board elections across America. Biologists have grown weary of seeing any statement they publish hinting at the limits of their theory being used against them by religious ideologues eager to demolish Darwinism by any means necessary and "prove" the literal veracity of the Good Book. So the biologists have simply learned to avoid making such statements. Indeed, they have assumed a habitually defensive—even self-policing—attitude. Rather than presenting the neo-Darwinian synthesis as a theory to be held provisionally and examined in the light of accumulating evidence, proponents treat it as proven fact—sometimes going so far as to regard it with a quasi-religious veneration. One result of this is an implicit requirement that graduate students adhere strictly to the received dogma. As Lynn Margulis puts it, ". . . neo-Darwinist fundamentals, derivative from the mechanistic life science world view, are taught as articles of true faith that require pledges of allegiance from graduate students and young faculty members. . . .[1] If an individual with ambition to study nature rejects neo-Darwinist biology in today's ambience, he becomes a threat to his own means of livelihood. . . ."[2]

A more immediate reason for the dominance of reductionism in university and corporate research labs has to do with money. Genetic engineering—based on the reductionist assumption that moving a discrete gene will produce a well-defined, controllable effect—promises commercial applications and potential profits. The Gaians' work usually doesn't. There was a time when Wall Street and the ivy-covered halls of academia were worlds apart; no more. With universities becoming increasingly dependent on corporate contracts to fund basic research, economic imperatives are beginning to shape scientific ideology. Given a choice between research that leads to dividends versus research that leads merely to a better understanding of the complexity of living things, universities face tremendous pressure to devote their resources to the former.

These two motives for the dominance of reductionism in modern biology—defensiveness against religious fundamentalism and the need to produce commercial applications in order to justify research—are fairly obvious. However, there may be still other, deeper influences at work. Biotech is narrowly the product of certain recent scientific discoveries, but from a larger perspective it is the inevitable result of fundamental transformations of thought that have shaped the recent history of our entire culture. Thus if we wish to understand biotech in its larger context—its meaning for our species and for the biosphere—we must understand, not only the perspective of the scientists who now control most university biology departments and corporate research programs, but also the currents of belief and assumption that condition our entire civilization.

Machine Mind, Computer Mind

Science does not exist in a vacuum. Scientists are human beings who are born, live, and die within cultural and economic circumstances that shape their view of the world. As it happens, modern genetic and evolutionary theories have emerged during a historical period that has come to be known as the industrial revolution, and it is only natural that the whirlwind of technological, economic, and social change that has characterized this period would have influenced the concurrent development of scientific thought.

People in every society and in every age tend to create a cosmology that reflects their needs and experiences, so it is hardly surprising that the machine age has seen the ascendancy of the idea that the world is itself a great machine. The machine metaphor first gained recognition toward the end of the Middle Ages; by the time Newton sought to derive the motions of celestial bodies from mechanical and mathematical principles, it was already beginning to dominate the thinking of the scientific elite throughout Europe.

Newton's universe was a great clock, with planetary orbits performing the function of gears. This model's success in predicting celestial motions led to the elevation of physics, astronomy, and mathematics to positions of preeminence among the sciences; from then on, researchers in other disciplines tried to explain every phenomenon they studied in terms of mathematically measurable, mechanical processes.

Charles Darwin regarded himself as a sort of Newton of biology, offering the world a quasi-mechanical theory of the origin and development of species. While the idea of nature-as-machine had roots, not just in the physics of Newton, but in the philosophies of Bacon, Hobbes, and Descartes as well, Darwin added an essential historical dimension to the discussion: Not only are all living organisms composed *solely* of insensate matter obeying physical laws, but they have been assembled over eons of time into their present functional combinations by a process that is random and purposeless.

Gradually, during the latter half of the 20th century, industrial processes began to be regulated more and more by electronic computational machines. As these machines have evolved from the Univac to the modern desktop computer linked via phone line to the Internet, society itself has become increasingly obsessed with data, information, and cybernetics (automatic control systems).

Along with the advent of the first useful electronic computers, the 1950s also heralded the discovery of the DNA double helix. Almost immediately, scientists began drawing comparisons. Soon the computer was the dominant metaphor for reconceptualizing the origin and development of species. Biologists began using the language of cybernetics to describe living systems, with comments like the following (by W. H. Thorpe) becoming commonplace: "The most important biological discovery of recent years is the discovery that the processes of life are directed by programmes . . . [and] that life is not merely programmed activity but self-programmed activity."[3]

Our hunter-gatherer ancestors experienced nature as an

anarchic community of spirit beings, plants, animals, weather phenomena, and celestial objects. Medieval Europeans, accustomed to a strict hierarchical social order, viewed nature as a Great Chain of creatures, all subordinate to humans (who were themselves subordinate to God and the Church), and all obeying an orderly and ranked set of obligations. With the dawning of the Industrial Age, nature was seen as a machine and a storehouse of resources. With the capitalist economic transformation of European society, nature also became an arena of fierce and unending competition, with only the fittest surviving. Now, in the Information Age, nature may be considered a set of programs and codes. All of these metaphors have no doubt been useful—comparing computer codes with DNA sequences, for example, has taught us a great deal about how genes actually work—but every metaphor is useful only up to a point. When we begin to see the reality as *nothing but* the metaphor, we run into trouble. Then we tend to overlook aspects of the reality that the metaphor cannot reflect.[4]

A Ghost in the Machine?

The machine metaphor (of which the computer metaphor is an extension) requires that the interchangeable parts that make up the living machine be inherently inert and without desire or aim. They are tossed about by impersonal external forces that can be described with precision using the language of mathematics. Darwin, in developing this metaphor, maintained that evolution was driven entirely by external causes, the environment pushing or pulling the inert organism in various directions. The cause of organic change, in Darwinian evolution, lies *outside* the organism.[5]

What is to be done, then, with evidence of inner purpose, such as that provided by Cairns's experiments, cited in the last chapter? In a strictly reductionist scheme, the idea that organisms deliberately pursue goals must be rejected, since "purpose" cannot be reduced to the laws of physics. Biologist Alex Novikoff assumes this extreme reductionist viewpoint when he writes, "Only when purpose was excluded from descriptions of all biological activity . . . could biological problems be properly formulated and analyzed."[6] This trend of thinking is traceable to Darwin himself, who suggested that while organisms evolve structures that look *as if* they were purposefully designed, and while living things behave *as if* they were consciously pursuing goals, this is actually an illusion; in reality, every organism—and every part of that organism—changes its form and behavior randomly; over time, those changes or behaviors that promote survival and the ability to reproduce will be favored. The *appearance* of inner purpose is thus reducible to chance variation plus natural selection.

However, the idea that organisms have no inner sense of purpose is contradicted by our own human experience. We each make plans, formulate goals, and pursue strategies routinely. And there is every indication that other creatures do the same, if perhaps not as consciously. The evidence is so persuasive that many biologists who otherwise subscribe to a reductionist-mechanist view are nevertheless forced to acknowledge some capacity of inner purpose on the part of organisms. For example, P. B. and J. S. Medawar, in their biology textbook *The Life Sciences,* write: "Purposiveness is one of the distinguishing characteristics of living things. *Of course* birds build nests in order to house their young and, equally obviously, the enlargement of a second kidney when the first is removed comes about to allow one kidney to do the work formerly done by two."[7] And cell biologist Edmund Sinnott has written: "Life is not aimless, nor are its actions at random. They are regulatory and either maintain a goal already achieved or move toward one which is yet to be realized. . . . [Every living thing exhibits] activity which tends toward the realization of a developmental pattern or goal. . . . Such teleology [or purpose], far from being unscientific, is implicit in the very nature of the organism."[8]

But if the existence of purpose in organisms is problematic for the purely mechanistic explanation of life—and for the more general philosophy of *materialism,* which holds that all observable phenomena are explainable as the results of material causes—consciousness is doubly so. Philosophers and theologians have long cited human thought, feelings, imagination, and reflection as proof of the existence of the soul and of a spiritual dimension beyond the physical. Understandably, reductionist and materialist science—which is at war with theistic philosophies featuring a non-physical God at the center of cosmos and creation—has therefore sought to find purely physical, chemical explanations for consciousness in humans and other creatures.

This effort has lately been rewarded by discoveries in the field of neurobiology. For many decades, studies of brain abnormalities or injuries suggested that if a person lost a certain ability following an injury or stroke, the damaged brain region must have been responsible for that ability. Through such studies, scientists have come to identify the occipital lobes as the area where visual information is processed, the cerebellum as the site governing muscle coordination, and so on. More recently, and perhaps even more convincingly, similar research has shown that damage to certain brain regions can produce specific, predictable personality changes. If a certain kind of brain damage can cause a normal person to become expressionless, unable even to "hear" the emotional content of music, isn't this proof that consciousness is *reducible to* brain anatomy, chemistry, and electrical activity, all preprogrammed by chains of chemicals within strands of DNA?

It might be, if there were no contradictory evidence—but there is. If consciousness were *nothing but* the firing of neurons, then we should expect it to be strictly localized within the cranium. How can we explain the voluminously documented research—much of it funded by the U.S. Department of Defense and the CIA—into the practice of "remote viewing," in which test subjects have repeatedly shown the ability to project their consciousness through time and space to retrieve information they could not possibly have obtained through any known physical means?[9]

There is good evidence for phenomena like clairvoyance, telepathy, and the healing power of prayer, but no scientifically "acceptable" explanation. Parapsychological phenomena have also been reported in animals (for example, in abundant anecdotes about pet dogs who "know" when their owners are about to return, even in the absence of routine schedules).[10] Such phenomena clearly suggest that there is some aspect of consciousness that *cannot* be easily reduced to the mechanical actions of known materials and forces.

Intriguingly, the most promising explanations so far for these transcendent capabilities of consciousness have come from physics—though not the mechanistic physics of Newton. Since the 1920s, quantum physicists have had to make peace with the fact that subatomic particles appear to behave in ways that confound our commonsense assumptions about causality and action at a distance. More than one quantum physicist has noted that these observations and the new theories required to explain them may tell us something about the human mind's apparent ability to overcome the bounds of space and time. Physicist Paul Davies captures the irony of this situation when he writes that ". . . physics, which led the way for all other sciences, is now moving towards a more accommodating view of mind, while the life sciences, following the path of last century's physics, are trying to abolish mind altogether."[11]

Rupert Sheldrake is one of the few biologists to have taken the phenomena of parapsychology seriously. He proposes that, in order adequately to account for them, neuroscientists may ultimately have to adopt the view that "the conscious self [has] . . . a reality which is not merely derivative from matter."[12] Sheldrake suggests that consciousness interacts first with motor fields associated with the brain and then, via the brain, with the rest of the physical body. Damage to the brain—an essential link connecting consciousness with the body—might then be expected to produce the sorts of memory, sensory and motor impairments, and personality changes that are actually observed.

To be fair to materialists, one must point out that their abhorrence of spiritual explanations for biological phenomena rests on a coherent philosophical foundation. Science seeks to understand natural phenomena by attributing them to comprehensible, observable causes. To accept what might be called mystical or metaphysical explanations for phenomena and events—that is, explanations that appeal to some cause that is immune to further analysis ("God did it"; "it's God's will")—is to short-circuit the entire scientific enterprise. Moreover, relying on such explanations can have limiting practical consequences. If, for example, an airliner crashes and we account for the event by saying that "it was God's will," we may have no motive to dig deeper and determine whether the design of the plane was faulty. Merely to say that "spirit" or "God" or "the inner self" directs evolution, or governs the expression of genes, or transmits instinct, explains very little and could prevent us from pursuing more fruitful lines of research.

Nevertheless, materialists and reductionists are sometimes guilty of relying on quasi-mystical explanations of their own. When confronted with the problem of explaining the origin of an obviously purposeful and profoundly complex organ such as the human eye, they are inclined to say, in effect, "Natural selection did it," without offering a convincing step-by-step scenario by which it could have done so, or evidence that it did.[13] Since no better material explanation is apparently available, it is assumed that whatever explanation is at hand—however obvious its shortcomings—*must be* true. Natural selection thus becomes an inscrutable, godlike agency capable of producing miracles.

At a certain point, *all* explanations become mystical, since they all eventually invoke causes that are immune to further analysis. For physicists, the terms *time, space, force, field,* and *charge* point to concepts that are so axiomatic that it is impossible to define them further except in terms of one another. Nevertheless, the goal of science is ultimately to reduce such undefinable terms to a minimum and to explain fully how the material universe comes into being and operates as a result of their interactions.

Gaians generally have no objection to this goal but suggest that, in its zeal to weed out irreducible or undefinable terms, mainstream science has gone too far, too fast and that the phenomena of life and consciousness have been glibly explained away. Even if life and consciousness depend for their manifestation on physical structures (like the human body) that can themselves be reduced to atoms, charged particles, and fundamental forces, these "higher" properties *may not* themselves be so reducible. The whole may, in fact, be greater than the sum of its parts.

The idea that evolution and the expression of genes are activities at least partly directed from within the organism—which can be thought of as a spiritual idea—may or may not turn out to be a fruitful explanation for the diversity of life; that will depend on whether the fundamental process involved can be further demonstrated, analyzed, and understood. If the notion helps fill in our understanding about biological processes, we should pursue it.

In any case, biology must somehow come to terms with the implications of Cairns's experiments. In addition, it must reconcile itself with the new physics. And it must begin to accept the hard evidence accumulating in the field of parapsychology, rather than ignoring it merely because it appears to contradict materialist dogma.

The Essence of the Sacred

Why is all of this relevant to the debates about cloning and genetic engineering? In the deployment of biotechnologies we are profoundly altering our relationship with the living world, so it is important to define what we think the living world is. Is it a complex heap of cybernetic chemical factories? Or is it a vast network of spirit-filled beings?

As we have seen, the choice is partly determined by cultural conditioning. For the scientist living in a modern industrial city, the living cell *is* a miniature cybernetic chemical factory. For the indigenous African or Australian or Native American shaman, life *is* the evidence of an indwelling spirit. In either case, the world view tends to be self-confirming as the result of a subtle culture-based psychological process that causes us to pay more attention to sensory data that confirm our worldview than to data that fundamentally challenge it.

But this postmodern, cultural-relativist way of looking at the construction of knowledge has its limits, as it implies that one worldview is just as good as another. Our choice of metaphors for the natural world may be culturally conditioned and therefore in some sense arbitrary, but on the basis of metaphors we may take actions that have profound consequences for ourselves and our environment.

Currently, those who are deploying biotechnologies (geneticists and corporate managers) tend to share one view of nature—the "cybernetic chemical factory" view, which is a fundamentally reductionistic, mechanistic, materialist perspective. Certainly, for the purposes of philosophical discussion, they have every right to such a view. However, when they propose actions that affect all of us, including many nonhuman species, shouldn't we also explore alternative lines of thought?

With regard to biotech issues, organismic, vitalist, and *even overtly spiritual views* about the nature of life deserve consideration. After all, the vast majority of the world's people hold some form of spiritual philosophy—whether it derives from a native, Earth-based spiritual tradition or one of the great world religions (Christianity, Buddhism, Islam, Judaism, Hinduism, or Taoism). Must we assume that all of these traditions are rooted in nothing more than superstition?

As previously noted, philosophical materialists find the spiritual world view distasteful because it introduces the idea of a nonmaterial cause for physical phenomena, leading ultimately to explanations that appear to boil down to some form of the statement, "God did it." But at least the spiritual view leaves open the door to the possibility that our explanations for biological phenomena are still incomplete in some fundamental way. To prematurely close that door might be a profound error. If we think we have essentially the whole picture of what life is and how it works, when in reality we have only a part of that picture; if our working philosophy systematically excludes certain kinds of evidence and certain kinds of explanations; and further, if we act on our philosophy in ways that have global repercussions, then we could be getting ourselves into serious trouble indeed.

A spiritual perspective, even in its weakest and most generalized form, would hold that present material explanations for biological and psychological realities are necessary but not sufficient. Something else must be taken into account. The exact nature of this "something else" may be no easier to define than words like *space, time,* or *charge.* But it is no less real. It may be some undiscovered field or force (as vitalists suggest), or a transcendent reality beyond the world of space and time (as proposed in various religious philosophies). On the other hand, this "something else" may turn out to be a pattern of organization (as predicted by the organicists). Living systems theorists or organicists, like Margulis and Capra, manage to avoid both reductionist and mechanistic thinking *and* the idea that there are fundamental, undiscovered, nonmaterial fields or forces that act from within living things and/or human consciousness. Organicists sometimes adopt spiritual terms to portray the ecological interconnectedness of living things. Capra, for instance, writes:

> When the concept of the human spirit is understood as the mode of consciousness in which the individual feels a sense of belonging, of connectedness, to the cosmos as

> a whole, it becomes clear that ecological awareness is spiritual in its deepest essence. It is, therefore, not surprising that the emerging new vision of reality based on deep ecological awareness is consistent with the so-called perennial philosophy of spiritual traditions, whether we talk about the spirituality of Christian mystics, that of the Buddhists, or the philosophy and cosmology underlying the Native American traditions.[14]

Most Gaians believe that the "something else" that is missing from reductionist descriptions of nature impinges on the physical world via living things, and is subjectively experienced by each living organism as its *self*. As Varela and Maturana point out, life can be defined as the capacity for *self-directed* motion or change.

The "self" that directs motion or change—whether we think of it in religious terms as a spirit being, a part of God, a reflection of God, or a creation of God; in vitalistic terms as entelechy or morphogenetic fields; or in organismic terms as the autopoietic organization of the whole—may depend for its manifestation on a complex network of physical structures and may in fact arise from that network. In any case, it is purposeful. It is an end in itself, and the organic constituents of its body are its means.

To use a spiritual term, it is very close to the core of what is meant by the term *sacred*.

The intuition, perception, or belief that other beings have a self with an interior experience comparable to one's own is the basis for ethics. When we feel compassion or love for another being, our inner self reaches out to that of the other, and its interests become ours. When we disregard the interests of other beings, we consequently diminish the manifestation of our own sacred essence.

All organisms are, in some sense, the subjects of their own experience and therefore not just objects to be manipulated without some degree of consideration. We see the capacity for self-directed motion or change in even the smallest bacterium. Its cell wall divides it (*self*) from its environment (*not-self*). It has rudimentary capacities for the perception of its world; it has ways of obtaining what it needs; it has its likes and dislikes. Even the bacterium has an interest in achieving its aims and not having those aims frustrated. Obviously, it is silly to think that the interests of a bacterium are ethically equivalent to those of a human. There is clearly a wide divergence in the degree of sentience in living things. Certainly we do not observe evidence in plants of the same kind or degree of sentience seen in complex animals. But the mere fact of their *aliveness* brings even plants within the sphere of our moral concern.

The Gaian view also recognizes that no individual self can exist apart from the web of life; all beings are interdependent. The natural world is a realm in which most organisms eat, and are eaten by, other organisms. The lion eats the gazelle; the same lion concurrently is host to microscopic parasites. While predatory behavior might appear to violate spiritual ethics, ultimately it serves spirituality by working to the benefit of all. Predator species and prey species do not compete with one another; their populations tend to be mutually stabilizing. In this sense nature is characterized far less by competition than by cooperation within and among species.[15]

The Gaian perspective is both holistic and ecological. It views nature as the cooperative, interdependent, self-directed activity of numberless creatures. It regards nature as the ultimate model of economy, cooperation, simplicity, beauty, and purpose. Ecological thinking implies a respect for wholes—not only whole organisms, but whole natural systems that arise from intertwined homeostatic relationships between and among living beings. And these irreducible larger wholes (ecosystems) deserve moral consideration too, far more of it perhaps than any individual organism.

If we are to take a moral and spiritual look at biotechnology, we must consider the Gaian account of the nature of life. Granted, such a view might seem to prejudice the outcome of our inquiry into biotechnology from the outset. After all, from a spiritual point of view, the terms *genetic engineering* and *biotechnology* are virtually oxymora.

However, we have yet to look at the nature of the techniques themselves and the specific moral arguments surrounding their use. Ethical questions are seldom black and white. Just what technologies are we talking about? How and in what instances are they being used? What good might come from them? And what are the motives for their development?

CHAPTER THREE

Biotech Basics

PERHAPS THE BROADEST possible definition for the word *biotechnology* is "the alteration of living things to produce useful products and processes." Thus defined, biotech is nothing new. Ever since the Neolithic revolution—roughly ten thousand years ago—humans have been breeding plants and animals to produce new strains with desirable traits. The patient efforts of countless generations of animal breeders have produced some amazing accomplishments—scores of breeds of dogs, horses, sheep, cattle, pigs, and goats with a tremendous range of novel or enhanced characteristics. Compare a Chihuahua with a Great Dane: both are members of a single species, and their obvious differences are mostly due to the work of dog breeders. Plant breeders have produced an even more impressive array of achievements, including hundreds of varieties of maize, rice, wheat, potatoes, apples, and other edible grains, fruits, and vegetables. Some food plants (like wheat and maize) have been so thoroughly domesticated, and for so long, that the original wild species from which they originated are difficult to identify.

The most basic breeding technique involves simply selecting individual animals or plants with a desirable characteristic and causing them to produce offspring; then the offspring in which the characteristic is most pronounced are again bred together. The breeder continues the process until the desired trait is maximized within a stable population.

Another technique breeders have used to alter plants (and, to a much lesser extent, animals) is hybridization—the production of offspring by crossbreeding members of different but closely related species. Many plant species, but almost no animal species, hybridize in nature. Even with human prompting, only a few animals produce true hybrids (the mule, the best-known example of a true animal hybrid, is the offspring of a male donkey and a female horse; like most animal hybrids, mules are usually reproductively sterile). However, the number of cultivated plant hybrids nearly equals the total number of plants in domestication, including thousands of varieties of wheat, corn, oats, tomatoes, squash, roses, tobacco, and many more. Luther Burbank, an early twentieth-century pioneer plant breeder, used hybridization and the selective planting of seeds to produce an unprecedented abundance of new varieties—more than two hundred varieties of fruit alone.

In this broad sense, biotechnology has brought many obvious benefits. In fact, it could be considered the necessary material basis of civilization: without domesticated grain crops, cities could never have developed and humankind might still be living primarily by hunting and gathering wild foods.

However, traditional biotechnology has always been confined within certain limits. Plant and animal breeders long ago learned that each species can be bred only so far in any given direction. It is possible to breed a dog that's as big as a sheep or as small as a skunk, but nobody can breed one that's as big as a horse or as small as a mouse. Each species appears to have natural genetic boundaries. As Luther Burbank put it:

> There are limits to the development possible, and these limits follow a law. . . . [It is the law] of the reversion to the Average. . . . Experiments carried on extensively have given us scientific proof of what we had already guessed by observation; namely, that plants and animals all tend to revert, in successive generations, toward a given mean or average. . . . In short, there is undoubtedly a pull toward the mean which keeps all living things within some more or less fixed limitation.[1]

Moreover, as natural limits are approached, negative side effects sometimes show up. For example, the Dachshund—a dog bred to have a long body and short legs—has a pronounced tendency to develop intervertebral disc disease; for more obscure reasons, Dalmatians have a genetic inclination to develop uric acid stones. When breeders make corn with bigger ears, or chickens that lay more eggs, they do so at the expense of the organism's overall viability. The strain produced is weakened in some respect and less resistant to environmental stresses (such as pests, diseases, or climate variation). Douglas Scott Falconer, former chairman of the Department of Genetics at the University of Edinburgh, sums up the situation this way:

> The improvements that have been made by selection in these [domesticated breeds] have clearly been accompanied by a reduction of fitness for life under natural conditions, and only the fact that domesticated animals and plants do not live under natural conditions has allowed these improvements to be made.[2]

The principal means of enhancing a domesticated plant or animal's survival ability is to control its environment—to systematically destroy pests and competitors and supply it with needed nutrients.

In the case of plants, as crops are bred further and further from their wild state, farmers must fight weeds and provide nutrients ever more intensively. Thus the "green revolution" of the 1960s and 1970s, which relied on new hybrid varieties to produce higher yields, also required greater reliance on petroleum-based pesticides, herbicides, and fertilizers. This chemical-based, industrial style of agriculture enabled farmers to increase their output (over the past fifty years, crop yields in the U.S. have risen on average by 1 to 2 percent per year), but it also produced far more environmental pollution. At the same time, it required far greater investment in equipment and chemicals, leading to increased debt, and thus driving many small-scale farmers out of business. Meanwhile, throughout the world a handful of green-revolution hybrids have replaced thousands of ancient plant varieties, leading to an erosion of genetic diversity in the global food supply.

Still, despite these limits and unintended consequences,

most people applaud the achievements of traditional biotechnology. Without it, we might not be able to feed the burgeoning world population. So why is there so much controversy about *new* biotechnologies that involve techniques like cloning and gene splicing? Both critics and supporters of biotech agree it is because they involve the manipulation of living things in fundamentally novel ways.

A definition of biotechnology in this new sense is "the alteration of *cells or DNA molecules* by means of splicing and cloning techniques to provide useful products or processes." With the new techniques, former barriers and limits are cast aside. While the transfer of genes between unrelated organisms does occur in nature (for example, through the action of viruses), the new biotech methods can produce genetic changes that would never occur in the wild or through even the most intensive breeding. Scientists can penetrate genetic barriers between species at will, isolating the DNA codes for given characteristics and inserting them into the genome of any organism they choose. They can mix the genetic material of species that could never be bred by traditional methods (bacteria and fish; humans and pigs; fireflies and tobacco plants). In principle, they can even avoid the tradeoffs of traditional breeding, such as overall lessening of an organism's viability as it is bred further in any desired direction.

The effect on human society of plant and animal domestication has been compared with that of the discovery of fire, or *pyrotechnology*. Our distant ancestors used fire primarily to warm themselves and to cook food; now we use it in a myriad of ways—to make new materials, to power automobiles and aircraft, to produce electricity. We use pyrotechnology *industrially.* Similarly, while our ancestors altered and used living things intuitively and in limited ways, we are now harnessing and changing them systematically and fundamentally, from the inside out. If pyrotechnology fired the industrial revolution in nineteenth-century Europe, then biotechnology represents a new industrial revolution, one in which life itself is both the fuel and the raw material.

The purpose of this chapter is to describe some of the basic processes involved in biotechnology. We've already discussed some philosophical and scientific objections to the assumptions on which it is based; toward the end of this chapter we'll see whether those objections are justified in practice. But first it is important to understand exactly what biotech is and what its practitioners are actually doing. What's all the fuss about?

Recombinant DNA

The historical roots of the modern biotech revolution go back to the 1600s, when Robert Hooke first saw individual cells through a primitive microscope. A second fundamental insight came in the 19th century, when Austrian monk Gregor Mendel discovered the laws of heredity—the statistical laws governing the ways characteristics are passed from one generation to the next. The ascendancy of molecular biology and the philosophy of genetic determinism in the American life sciences, resulting from the Rockefeller Foundation's patient funding efforts and political advocacy, set the stage. But two profound discoveries in the 1950s and 1960s—Watson and Crick's unraveling of the double helix and Har Gobind Khorana and Marshall Nirenberg's cracking of the genetic code—ultimately led to a virtual explosion of breakthroughs in biotech.

In 1968 two Swedish scientists, Dr. Torbjorn O. Caspersson and Dr. Lore Zech, found a chemical capable of staining one of the four bases of the DNA ladder. When stained and placed under ultraviolet light, a chromosome would glow in a pattern of bright and dim spots reflecting high and low concentrations of base G. Soon other stains were developed and the technique was used to map genes and connect them with specific traits and disorders.

The ability to identify genes soon led to the discovery of a powerful new method of manipulating them. In 1973 biologists Stanley Cohen of Stanford University and Herbert Boyer of the University of California chemically isolated strips of DNA from two different species, combined them, and inserted them into a host cell. Known as *recombinant DNA* or *gene splicing*, this technique has become the basis for genetic engineering and is perhaps the most dramatic development in the biotech revolution to date.

Right now, I'm typing into a computer, using a word-processing program that allows me to cut pieces of text and splice them together in any order I choose. The procedure that Cohen and Boyer pioneered gives biotech scientists a similar power: using recombinant DNA techniques, pieces of DNA can be cut and spliced together to create new statements in the language of genes.

Suppose we wish to add a few toad genes to a bacterium

(this is, in fact, exactly what Cohen and Boyer did). The first step is to cut a long filament of toad DNA into fragments that hold the specific genes of interest. This is done by first obtaining toad DNA, using a centrifuge and chemicals to break open the membranes of previously harvested toad cells. We then introduce to the DNA a tiny amount of a bacterial biochemical known as a *restriction enzyme*. Each restriction enzyme (more than eight hundred of which have been discovered, with many now being produced commercially) snips the double-stranded DNA molecule at a particular sequence of nucleotides. The cut usually leaves short, single-strand bits of DNA dangling from both fragments. These "sticky ends" of unpaired bases can then be used to bond the cut pieces to other DNA fragments that have complementary dangling strands.

In order to transfer the toad genes to living bacterial cells we need a *vector,* which is a naturally occurring transmitting agent—usually a virus (one that has been manipulated to transport genetic material without causing disease) or a plasmid (a small circle of DNA found in some bacteria). Let's suppose we've decided to use a plasmid vector. First, we chemically remove the desired plasmids from a culture of bacteria cells. Then we cut them with the same restriction enzyme we used to slice up the donor DNA, so as to expose complementary sticky ends. When we mix the plasmids and the toad DNA, the sticky ends join up. Since the DNA code is universal, it doesn't matter whether the DNA fragments we're seeking to join were taken from a bacterium, a mouse, or a human; as long as the sticky ends match, the fragments will stitch themselves together. However, the connection is fairly weak until sealed with another enzyme called a *ligase.*

The result of the process so far is a mixture containing plasmid circles with added toad genes, along with a lot of leftover donor and plasmid DNA. When we add bacterial cells to the mix, the bacteria absorb much of the material. All that's left is to separate the bacteria with the toad DNA from the rest. Knowing that we would need to do this at the end of the process, we will have begun by selecting plasmids that contain a gene for resistance to a particular antibiotic. That way, we can find which bacteria have taken up the plasmid by exposing them to the antibiotic: the ones that survive carry the plasmid. We can then use a different antibiotic test to determine which of the bacterial cells containing the plasmids also include toad genes.

This general procedure is now commonplace in making genetically engineered bacteria, and it is industrially useful for producing large batches of biochemicals such as insulin or interferon. Just splice a gene that codes for the production of the desired biochemical (along with the DNA "switch" that turns the gene on) into a vector plasmid, add the plasmid to a culture of bacteria, and separate the bacteria that carry the new genes. Then all that remains is to grow vats of the altered bacteria, siphon off the desired biochemical, and test it for purity.

The range of applications for the technique is virtually endless. Genes can even be engineered to help make jeans. Natural sources for indigo, the dye used in denim, include certain mollusks and the fermented leaves of the European woad plant or the Asian indigo plant. In 1883 chemists discovered a synthetic process to produce the dye using coal tar, but the process requires hazardous reagents and produces toxic by-products. Exactly a century later, biotech researchers found that *E. coli* could be genetically programmed to produce indigo from glucose, a simple sugar. As the new procedure comes into general use, the environmental problems resulting from coal-tar production of indigo will be eliminated.

Meanwhile, your next pair of blue jeans could be a biotech product in more than one respect: not only the indigo dye, but also the cotton fiber from which they're made may be genetically altered.

Transgenic Organisms

Bacteria are easy to engineer. They have no nuclei, so their DNA floats freely throughout the cell. They soak up extra plasmids as effortlessly as teenagers absorb junk food. The process of splicing bacterial DNA is so simple that eighth graders in some middle schools now routinely practice it in biology class.

But suppose we want to genetically engineer a cotton plant or a mouse. Suddenly a host of new problems crops up. Clearly, in the case of the mouse, the best way to do it would be to alter the genes in a fertilized egg cell, then implant the embryo in the uterus of a female mouse. But how can we get the DNA to enter the egg's nucleus? And even if we solve that problem, how can we get the gene to go to where we want it on the mouse genome, so that it will be expressed in the animal?

In 1983 Ralph Brinster of the University of Pennsylvania

Veterinary School became the first to surmount these hurdles, successfully inserting human growth hormone genes into mouse embryos. In Brinster's experiments, very few genes actually went where they were intended, so that scores of mice had to be implanted with engineered eggs before one gave birth to a transgenic offspring. But the eventual result was impressive—a "super mouse" that grew twice as fast and nearly twice as large as ordinary mice. That strain of mice with human growth genes still exists, with the new characteristic permanently incorporated into its genetic makeup.

Inserting foreign genes into fertilized eggs is a demanding process. Early researchers had to spend countless hours sitting hunched over microscopes, carefully pushing a syringe containing the donor DNA into tiny embryo cells. The experiment almost never succeeded, since only about one embryo in five even survived the injection, only one in a hundred took up the added gene, and not all of those that maintained it expressed it in all their subsequently developed cells. New techniques have improved the success rate and reduced the tedious labor, but the genetic engineering of large animals is still a matter for serious professionals only.

Of course, the production of a bigger mouse would not, in itself, justify so much trouble. But there are other, more interesting and commercially promising uses for the techniques Brinster and others have developed. For example: meet Nancy, a transgenic sheep who produces human alpha-1-antitrypsin (AAT) in her milk. AAT is a chemical normally present in blood serum that helps the microscopic air sacs in the lungs inflate properly. People without AAT develop an inherited form of emphysema, and donated blood doesn't yield enough AAT to treat the twenty thousand people afflicted with the ailment. A herd of transgenic sheep like Nancy can easily supply the needs of all these people.

Nancy was the first mammal engineered to produce a commercially useful biochemical in its milk. But many others have followed in her tracks. Today, transgenic pigs provide hemoglobin blood substitute; rats produce the human growth hormone to treat pituitary dwarfism; sheep supply the protein needed to clot the blood of hemophiliacs; cows produce lactoferrin, an iron-building milk protein added to infant formula to prevent bacterial infections; and rabbits make erythropoietin, used to treat dialysis-induced anemia.

Transgenic animals are primarily interesting to scientists for their ability to "pharm" useful human proteins in their milk. But sometimes biotechnicians have other goals in mind. In 1984 British researchers Steen Willadsen and Carole Fehilly fused embryo cells from goats and sheep to produce a sheep-goat chimera—the first genetic blending of two unrelated animal species in history. Willadsen and Fehilly made fourteen sheep-goats (or "geep") in all, along with several sheep-cows—which looked like sheep with Holstein-like black spots. The point of the exercise was not just to satisfy idle curiosity but to see whether the species barrier could be broken in pregnancy. The experiments showed that it would be possible to breed animals of endangered species in laboratories by implanting fertilized embryos of the endangered animals into surrogate mothers of related species. Still other applications easily come to mind: dairy farmers would like to have all-female herds; why not make transgenic bulls who could father only females? The reverse could be done for beef producers, who would prefer to have only male calves.

Like animals, transgenic plants can be programmed to produce pharmaceutical chemicals. Genetically altered soybeans are especially efficient in producing pharmaceutical proteins (such as cytokines—immune system biochemicals), since they are 40 to 45 percent protein to begin with.

However, the purpose of most plant bioengineering is to produce food and fiber with desirable qualities or crops that can be grown more efficiently. For example, researchers have developed a transgenic cotton that grows in the presence of the herbicide bromoxynil (sold by Rhone Poulenc, a French biotech chemical company, under the trade name Buctril). Cotton farmers were previously unable to use bromoxynil to kill ragweed, cockleburs, and other weeds because it killed cotton too; but with herbicide-resistant transgenic cotton, they are free to spray Buctril to keep weeds down thereby producing larger yields. Other transgenic plants that have already come to market include a tomato that ripens slowly (facilitating shipping); herbicide-resistant corn, canola, flax, and soybeans; insect-resistant potatoes; and a virus-resistant papaya. Others currently under development include bioengineered versions of alfalfa, apples, asparagus, barley, beets, broccoli, carrots, cauliflower, chestnuts, cucumbers, peanuts, peppers, rice, squash, strawberries, walnuts, watermelons, and wheat.

Cloning

One of the challenges of genetically engineering multicelled organisms is finding a way to insert DNA into cells (remember, plasmids can only be used with bacteria). Some of the methods now used involve *liposomes* (fatty bubbles that are easily absorbed by cell membranes and that can carry genetic cargo); *electroporation* (in which a brief jolt of electricity opens transient holes in cell membranes, permitting the entry of foreign DNA); and *particle bombardment* (in which a gun-like device shoots tiny gold or tungsten particles, coated with foreign DNA, through the cell wall).

The transgenic engineering of plants is especially challenging, because their cell walls are difficult to penetrate. However, in other respects plants are much easier to engineer than animals. Plant cells, unlike adult animal cells, have the potential to express any of their genes and thus repeat the developmental process from single cell to mature plant. Gardeners are familiar with this property. Take a cutting from an African violet and put its lower half into a moist growth medium. Then, when roots begin to form, transplant it—and voila! you have another African violet genetically identical to the first.

Your new African violet is actually a clone. Yes, home gardeners have been cloning for decades. Cloning even occurs naturally, with no human intervention at all. Identical twins are clones of each other, and cell cloning occurs each time a cell divides in two. Cancer cells, which can divide indefinitely (unlike other cells, which can continue dividing for only a limited number of generations), are cloning champions. This quality makes them useful to biotechnologists, who can join them to other cells to mass-produce valuable proteins, such as monoclonal antibodies.

The deliberate cloning of animals is a considerably more complicated affair. As long ago as 1938, scientists began trying to clone animals by removing an egg cell's nucleus, replacing it with the nucleus from another cell, then implanting the egg in a surrogate mother. It wasn't until the 1970s that frogs were successfully cloned by this method. However, none of the frogs developed past the tadpole stage. The problem the researchers encountered was that soon after the fertilized egg cells (zygotes) of animals begin to divide, they also begin to differentiate—forming into specialized organs such as skin, lungs, brain, or heart. Insert the nucleus from a differentiated cell into a host egg cell and not much happens. The nucleus is already programmed to express some of its genes and not others, so the body's entire blueprint can no longer be reproduced.

Later researchers avoided this problem by cloning only cells from embryos—usually clusters consisting of no more than eight cells. In some cases, embryo cells haven't yet started to differentiate. The process occurs at different rates in different species: mouse embryos start differentiating almost immediately, making them difficult—though not impossible—to clone; while bovine, sheep, monkey, and human embryos reach the eight-celled stage before differentiating, thus making the process easier.

Cloning from adult cells was considered impossible until a method of reprogramming them to express their entire repertoire of genes was developed. A Scottish team headed by Ian Wilmut made this breakthrough in 1996, using cells taken from the udder of an adult sheep. They froze the cells, thawed them, and starved them for five days in a growth medium containing only minimal nutrients. This procedure set the cells' developmental "clocks" back to zero, where they could be synchronized with that of the host egg cell. The end result of that experiment was Dolly the sheep, the first mammal cloned from an adult cell.

Dolly was soon on the front page of *The New York Times* and the cover of *Time* magazine. She was "the most important news story of the past two or three decades," according to James Fallows, editor of *The New Yorker.*[3] Wilmut had shown that adult mammals could be cloned and the implication was clear: humans could be cloned, too. Sci-fi scenarios of mad scientists producing carbon copies of Hitler or Jesus were now, in principle, realizable—assuming one could obtain a complete, undamaged DNA sample from either of those individuals. For more than a decade, leading scientists had insisted that the cloning of adult animals was out of the question. Suddenly it seemed as though the possibilities were limitless.

To be sure, the cloning of Dolly was a difficult process. She was the sole successful result of fusing adult cell nuclei with 277 different eggs. Even after Dolly, researchers were still a long way from human cloning; with the low success rate of Wilmut's technique, there would be an unacceptably high probability of miscarriages and birth defects if it were applied to humans. However, in the following months researchers in Hawaii and England refined the process, raising the success rate so significantly that use of the technique with humans now seems inevitable.

However, Wilmut's goal was never to clone people; in fact,

he has publicly stated his opposition to that use of the procedure. For Wilmut and nearly all the other scientists researching animal cloning, the goal is simply to create an unlimited source of animals with identical characteristics—monkeys for medical research, prized cattle for beef or milk, or even champion race horses. In addition, endangered species could be salvaged and transgenic animals bred to produce low-cost pharmaceuticals in their milk could be cloned to insure genetic uniformity.

The Biotech Workshop

In twenty-five years, biotech has gone from cutting-edge research to routine commercial application. Today, machines do much of the everyday lab work involved in identifying and engineering genes. For example, most agricultural, medical, and forensics biotech labs now use a PCR machine—a $7,000 unit about the size of a large toaster oven—to clone multiple copies of a given DNA sample. PCR stands for *polymerase chain reaction*; the machine uses heat to unzip the DNA molecule into two strands, then cools them and uses an enzyme called a polymerase to build up complementary strands. Then the heating, cooling, and enzymatic cycle begins again. Suppose you have a single DNA fragment and need a million or so identical copies: the PCR machine can deliver them in a couple of hours.

Another machine can assemble a couple of dozen nucleotides in any sequence required, producing synthetic single strands of DNA. These strands can then be used as probes to locate specific genes on chromosomes. If made radioactive and added to an unzipped DNA sample, the synthetic strands will pair with corresponding genes in the sample. Then, when the sample is X-rayed, the radioactive strand will show up as a dark spot on the film, revealing the location of the gene in question. This technique is called a DNA probe; it's the method of choice if you want to confirm the presence of recombinant DNA in a bacterial culture, find a disease-causing gene in a person's chromosomes, compare DNA "fingerprints" in order to help solve a crime, or screen potential employees for undesirable genetic traits.

PCR machines and nucleotide sequencers, along with other biotech lab equipment—new and used—can be found for sale on the Internet or in trade journals, and a basic biotech lab can be set up for as little as twenty to fifty thousand dollars (depending on its purpose).

In some respects human cloning represents merely the honing of basic procedures that are commonplace at approximately three hundred assisted-reproduction clinics nationwide. Through in vitro fertilization (IVF), ripe eggs are removed from a woman's ovaries and mixed with sperm, the resultant embryos are incubated, and the healthiest are selected and transferred back to the mother's uterus. IVF already entails many of the egg-and-embryo manipulation techniques that will be required for the cloning and transgenic engineering of humans; the biggest hurdle is finding more ways to increase the success rate in nuclear implantation.

While technicians already have many of the biotech basics well in hand, much research is still needed in order to realize the promise of the new science. For example, the ability to screen human chromosomes for all known genetic disorders will require much more knowledge of the estimated eighty thousand genes contained in the entire human genome. This knowledge is rapidly being acquired by a government-funded program known as the Human Genome Project, which was begun in 1987. Researchers, contracted by the U.S. Department of Energy and the National Institutes of Health to work on this project, hope to have read all three billion letters of the complete human DNA code by the year 2005 and to have gained an understanding of the functions of most of the genes it contains. At present, of the few thousand genes whose locations have been mapped, the functions of only a few hundred are understood; and in most cases we don't know how those functions are carried out, how the genes in question are expressed, or how they work in combination with other genes.

With the knowledge gained from the Human Genome Project, researchers hope to expand their use of gene therapy—compensating for, or even correcting, genetic defects at the molecular level. Gene therapy, in its current early stage of application, involves either adding normal genes to a patient's DNA to produce a biochemical the patient lacks, or obstructing the effects of genes that cause disease. But eventually biotech promoters hope to replace damaged genes at specific target sites on the chromosome, in both adults and embryos. At that point, it would be possible to produce "designer children" immune from specific diseases, or carrying specific genes to confer intelligence, good looks, or other desirable qualities.

Plenty of other applications are still in the realm of science fiction—such as the ability to rejuvenate "old" cells to make us young again or to replace damaged organs with human-cloned or transgenic-animal organ transplants. Despite the remarkable accomplishments of the past quarter century, including the introduction of hundreds of genetically altered medical and agricultural products, we are still at the earliest stages of the new technological revolution.

See—It Works! . . . Or Does It?

Geep. Cloned mice. Genetically engineered jeans. All of these seem to support the effectiveness of the mechanist point of view. *It works*, after all. Engineer a gene and a desired result appears. Insert a human growth hormone gene in a mouse and you get a giant mouse. Insert a gene for pesticide resistance into a corn plant and you get pesticide-resistant corn. Clone a sheep and you get another sheep nearly identical to the first. Surely these spectacular results are ample confirmation not only of the technology's value but also of the validity of its underlying theories and assumptions. The Central Dogma of molecular biology predicts a simple, one-to-one correlation between genes and traits. Isn't that what we are seeing?

In some instances, yes. Critics of genetic determinism do not deny the relationship between genes and traits; however, they point out that this relationship is complex. They predict that, along with the spectacular successes of genetic engineering, a host of unexpected side effects will appear—some with potentially horrific risks to the environment and human health. And current events are already confirming that prediction.

Take, for example, the story of an attempt to redesign the common soil bacterium *Klebsiella planticola*. In the early 1990s, German scientists engineered a version of *Klebsiella* that could digest wood chips. Lumber companies would be able to use the bacteria to break down wood waste; as by-products, the process would yield alcohol and a slimy sludge. The alcohol could be sold as fuel, the sludge as fertilizer. However, the transgenic *Klebsiella* could not be tested in Germany because of widespread public opposition there to biotech research. The German researchers appealed to the U.S. Environmental Protection Agency (EPA) to conduct the tests, which were arranged in Oregon. Experimenters grew colonies of bacteria in sand and examined the gases they released, and on this basis the EPA decided that the bacteria were safe. But a soil microbiologist at Oregon State University, Elaine Ingham, was doubtful; she proposed growing *Klebsiella* in pots containing soil and communities of common soil organisms; she also suggested planting wheat in the pots in order to test for possible effects on plants.

The results were disturbing: most of the wheat died. Moreover, Ingham found that in different kinds of soil the genetically modified bacteria killed the plants in two distinct ways. In one case, it caused roundworms to attack the wheat plants; in the other soils, it killed the plants by another means that is still not understood.[4]

According to Philip Regal, "This is a very serious issue, because the data show that this genetically engineered organism was very competitive and it probably would have spread. Let's say you put it in wheat fields: it would spread and kill the wheat, and it would not be safe to grow wheat there again. But in the worst-case scenario it doesn't stop at the edge of the wheat field; it just keeps going into the forest beyond the field and starts marching down the Pacific coast, destroying all the forests of Washington, Oregon, and northern California, turning them to slime. If that happened it would be an incredible disaster. We just have no way of knowing where the thing would stop."

The point is not just that genetic engineering produces some unanticipated problems. *Every* new technology produces new problems. But with genetic engineering the "side effects" are pervasive because the technology's conceptual basis is systematically flawed; given our imperfect understanding of living organisms, tinkering with genes will *inevitably* lead to unforeseen outcomes. Further, since the objects of manipulation are alive and can reproduce and spread their altered genes throughout the rest of the biosphere, the harm cannot easily be recalled or undone. The collateral ecological damage may be overwhelming.

Molecular biologist and lawyer Margaret Mellon, Agriculture and Biotechnology Program Director for the Union of Concerned Scientists, told me that, with regard to genetic pollution, "It will happen. No doubt about it. I think it will have effects that for the most part we will not see. In most cases it will advantage one little wild plant over another, and the overall complexion of the natural world will not be affected. But in some small percentage of cases, I would guess, it will have an impact that we care about, that will be unmistakable, and that, had we known about it, we would have chosen to avoid. I think the early instances are likely to be in the realm of agriculture. If you move glyphosate resistance genes into crops, those genes will make their way into weeds, and those weeds will no longer be treatable with glyphosate. And that will affect the choices of herbicides that farmers make. In some cases, because glyphosate actually is better than other herbicides that people want to use, it means that we're going to be using more dangerous herbicides. It's going to change the suite of choices that farmers have. And that is something I think we want to avoid. I think we would be far better off to rethink the way we do agriculture and to reserve glyphosate—which is a relatively better herbicide than others—for where we really need it and try to develop an agricultural system that needs to use it very rarely."

Then there is the problem of human health risks from genetically engineered foods. If a single gene can code for more than one trait, and if its effects depend partly on contextual factors (e.g., its place on the genome), this means that it will not always be possible to predict the exact effects that will ensue from inserting a particular foreign gene into a given organism. We may, for example, insert a gene for nutritional enhancement into a soybean plant but later find that this single, discrete genetic alteration has unanticipated side effects, such as causing an allergic reaction in many people who eat the soybeans.

This is, in fact, what happened when the seed company Pioneer Hybrid tried to transfer from Brazil nuts to soybeans a gene that codes for producing methionate, a nutrient often added to animal feed. The product was moving quickly toward the market and, in an effort to deal with possible regulatory issues, representatives of Pioneer Hybrid, the USDA, and the FDA sat down together. Government scientists knew that Brazil nuts can be highly allergenic and that some people can die from eating them. Pioneer Hybrid representatives argued that the chances of transferring the Brazil nut's allergenicity to the soybeans through a single gene were infinitesimally small. One government scientist present at the meeting expressed doubt and offered to perform an experiment. He found that the allergenicity had indeed been transferred, and a larger study later confirmed that result.[5]

It would be a mistake to assume that these examples are freak accidents in an otherwise well-understood process. As the *New York Times* reported, when Monsanto was developing its "New Leaf Superior" potatoes, which contain a gene coding for the production of a natural pesticide, scientists had to introduce the gene into thousands of potatoes in order to obtain the result it wanted because the gene usually ended up in the "wrong" place on the genome, resulting in a plant that either did not have the desired properties or that was an outright monstrosity. David Stark, co-director of Monsanto's potato subsidiary, Naturemark, is quoted by the *Times* as saying, "There's still a lot we don't understand about gene expression."[6]

Each individual instance of a genetically engineered food plant poses the risk of a novel or unexpected health problem. Philip Regal explained the situation to me this way: "All plants have biochemical pathways that can produce chemicals that can be toxic to animals. Often it's a defensive strategy on the part of plants. Sometimes they kill insects; sometimes they make the insects sterile or reproductively incompetent. There are whole categories of plant-produced compounds that can influence our physiology. We've learned to not eat plants that are going to kill us or make us sick. Our traditional foods are not in those categories. Either those chemicals are not present in our traditional foods, or they're present at such low concentrations that it's not a problem.

"But with genetic engineering, it's my concern that if you upset those biochemical pathways and start getting odd variants of these bioactive compounds, you will affect human health in some cases. It's like Russian roulette. Every time you genetically engineer something you're trying out a new chamber in the gun. Every project could be different: you could genetically engineer tomatoes with one insecticide in them and it might be okay, but then the next one might upset the biochemical pathways. We have to know in each case. I know this is a lot to ask for from the industry's point of view; it involves a tremendous amount of testing. But from a scientific point of view it's really necessary to treat each case as a new chamber in the gun."

At present, studies on the health effects of genetically engineered foods are not required by government agencies and are not being undertaken by the industry.

Yes, gene technologies enable us to do some amazing things. We can clone a mouse from a single adult cell, engineer a potato to resist pests, or redesign a cow to produce valuable pharmaceuticals in its milk. Science has given us a new and immensely powerful tool—but it is one whose implications and dangers we only vaguely understand.

One might therefore assume that the attitude of the technology's designers, managers, and regulators would be one of extreme caution—but quite the opposite is true. Despite some disturbing early experimental results (including the *Klebsiella* and Brazil nut-soybean disasters) and objections from ecologists, health experts, and concerned citizens, commercial applications of biotechnology are moving full speed ahead. Literally thousands of biotech products (seeds, food plants, pharmaceuticals, and so on) are moving quickly to market. Most Americans already consume genetically altered foods every day.

Why is caution apparently being thrown to the wind? What's driving the rush to exploit our new-found power over the very basis of life—before we have even begun to comprehend its meaning or debate its wisdom? Some of the impetus may arise from the mechanist-reductionist assumptions still held by powerful factions within the scientific community. However, as we will see in the next chapter, the profit motive may play a more central role.

CHAPTER FOUR

Patents and Profits

THE ABILITIES TO identify, map, and engineer genes and to clone plants and animals have given geneticists a set of basic tools that can be used in a wide variety of ways. The imagination runs wild with possible applications in the fields of medicine, agriculture, industrial processes and materials, environmental remediation, and biowarfare.

But putting basic biotech discoveries to work requires investments, industrial resources, and organized teams of employees. And, given the capitalist organization of our society, none of these can be mobilized without the expectation of profit.

A discussion of corporations and patents might appear out of place in a book on the spiritual and moral issues of genetic engineering. However, a little deeper reflection shows why these subjects are essential to understanding many of the moral concerns surrounding the new technology. The development of biotech is taking place, not in a vacuum, but within a legal and economic context which shapes it and which it increasingly shapes in turn.

Biotechnology is not "pure" science in the same sense as is the discovery of a new galaxy or a new principle in physics—

knowledge acquired for its own sake or solely for the sake of benefiting humanity. The story of biotech is not just about research and discoveries; it is also about money, and it is about turning genes—the heritage of millions of years of evolution—into the "intellectual property" of corporations. This privatization of the chemical basis of life has profound ethical implications.

Owning Ideas

Commercial biotech was born in 1976 in San Francisco when researcher Herbert Boyer (coinventor of genetic engineering) and businessman Robert Swanson formed a new company called Genentech, raising $35 million in a public stock offering. At the time, Genentech had yet to introduce a single product. When the nation's second biotech firm, Cetus Corporation, went public in 1981, it set a Wall Street record for the largest sum ever raised in an initial public offering. The financial community was clearly euphoric about the new technology's potential.

However, commercial returns were slow in coming. Bringing a biotech application to market typically requires many years and millions of dollars, so investments could not be recouped quickly. Some biotech products have met with staunch public resistance, and others have not lived up to initial expectations. The momentum of the entire industry was tempered by the three-year bear market of the early 1990s. And during the bull market of the late 1990s, while other stocks soared, most biotech stocks were treading water. Nevertheless, investments in research have accelerated, along with the expectation that eventually the thousands of new products making their way to market will result in huge profits.

The commercial exploitation of these new products hinges to a large degree on patent protection—the ability to claim temporary exclusive ownership over a product or process and to charge a royalty to others for the right to make or use it. This was true of Edison's light bulb, and it's true today of everything from software programs to paint thinners. Often, if you can't patent an idea or invention, it's worthless.

However, unlike light bulbs or computer programs, many of the "products" of biotechnology present a novel problem: they're alive—or, in the case of DNA sequences, they are fundamental aspects of the universal and primordial heritage of all living things on the planet. Is it ethical—even temporarily—to claim ownership not just of a particular cow, for example, but of a genetically identifiable kind of cow? Should we patent living things and their genes?

Clearly, the U.S. Patent and Trademark Office (PTO) thinks so; it has issued patents for genes, bioengineered life forms, and biotech processes. But upon these patents hinges a controversy with profound implications. To appreciate those implications, we need to know something about the current patent system itself and how it evolved.

The first "letters patents" were issued in the sixteenth century by European monarchs as licenses to conquer non-European lands and peoples. Soon, the same legal principle was being applied to inventions. The English Parliament passed the first patent statute in 1623; in the United States, Congress passed the first U.S. Patent Law in 1790 and created the Patent and Trademark Office in 1836. The aim of these early statutes was to compensate individual inventors for their work. President Lincoln declared in 1860 that "the introduction of patent laws added the fuel of interest to the fire of genius" by conferring the protection of monopoly to the "true inventor" of a novel idea, thus directly rewarding creative effort.

However, in the late nineteenth and early twentieth centuries, the process of invention began to be increasingly corporatized. Individual innovators were giving way to teams of researchers hired by large companies. (Thomas Edison himself was a leader in this trend, hiring scores of other inventors to work in teams toward the solution of technical problems.) In the pursuit of monopolies within their industries, corporations harassed independent inventors with lawsuits, forcing them to join the system or go bankrupt defending their patents in court. As individual scientists became corporate team players, they were required to sign agreements assigning ownership of their innovations to their corporate employers. In 1885 twelve percent of patents were issued to corporations. By 1950 the proportion had surpassed 75 percent.

While scientists and inventors were increasingly acting as servants of corporations, science itself was becoming a tool for the corporate management of society. As one observer told the Senate Patent Committee in 1949, "Today . . . the patent system adds another instrument of control to the well-stocked arsenal of monopoly interests. . . . It is the corporations, not their scientists, that are the beneficiaries of patent privileges."[1] In the same year, Robert Lynd, a contributor to the Herbert Hoover-sponsored study of *Recent Social Trends*, wrote: "The problem we face today is that, in an era that increasingly lives by science and technology, business control over science and its application to human needs gives to private business effective control over all the institutions of democracy, including the state itself."[2]

At the same time, corporations were seeking to socialize their costs—inducing government to pay for pollution cleanup, transportation infrastructure, and education of employees—while still funneling profits to investors. Basic research costs were often subsidized by universities and government-funded institutes. Sometimes corporations offered lucrative contracts to lure successful university researchers into private labs, on condition that they bring their patents with them; other times, companies partially funded university or institute research, with the understanding that forthcoming patents would be assigned to the company. This latter practice was eventually formalized in the Technology Transfer Act of 1986, which requires federal laboratories to actively seek opportunities to give or license patents developed at public expense to private corporations.

In an earlier era, patent protections were widely regarded as essential to scientific progress. But during the past quarter century, with the growth in size and power of transnational corporations, critics have begun to question whether patents in effect actually impede research and discovery. Studies such as Leonard Reich's "The Making of American Industrial Research" (1985) suggest that today patents are used primarily to stifle competition, blocking other firms from entry into the market.[3] Moreover, patenting tends to promote an atmosphere of secrecy within and around research laboratories, preventing scientists from openly sharing information. For example, a cancer researcher may be constrained from sharing preliminary findings with researchers at a different laboratory in order to prevent the divulgence of a key secret that might enable the other company to obtain a patent first. Many discoveries now showing up in patent files would formerly have been published in science journals.

Controversies about the patent system, the privatization of knowledge, and the corporate domination of society are significant in and of themselves, as we contemplate the future of democracy in an age of science and technology. But when the very stuff of life becomes the subject of corporate research and patent ownership, an even more crucial set of moral quandaries arises.

Patenting Potatoes, Pigs, and People

The idea of patenting life forms is not entirely new. The first such patent was issued in France in 1873 to Louis Pasteur for a type of yeast he developed for use in industrial processes. In the United States, the Plant Patent Act of 1930 gave growers the right to patent new plant variants developed asexually. This right was expanded in 1970 by the Plant Variety Protection Act, covering new varieties of sexually produced plants but excluding their seeds. At the time, few government officials could envision the possibility of creating a genetically engineered species with a commercial application.

Yet only a year later, that possibility had been realized. Ananda Chakrabarty, then an employee of General Electric Company (GE), applied for a patent on a genetically engineered bacterium designed to digest oil slicks at sea. The PTO at first rejected the request. Chakrabarty and GE appealed to the Court of Customs and Patent Appeals, which ruled that "the fact that the organisms . . . are alive [is] without legal significance" and that the patented bacteria were "more akin to inanimate chemical compositions such as reactants, reagents, and catalysts, than [they were] to horses and honeybees or raspberries and roses."[4]

The PTO appealed the decision to the Supreme Court and was joined in the case by the People's Business Commission (soon renamed The Foundation on Economic Trends)—an advocacy group led by Jeremy Rifkin, the most visible, eloquent, and persistent of biotech critics, whom industry-slanted articles invariably dismiss as a "Luddite" or a "gadfly." Rifkin's associate, Ted Howard, argued before the court that if the patent were upheld, then "manufactured life—high and low—will have been categorized as less than life, as nothing but common chemicals."[5] By a margin of five to four, the justices ruled in favor of GE and Chakrabarty. The Court stated that their decision should be narrowly construed and was not meant to apply to the larger social issues surrounding the engineering of life. Almost immediately, however, chemical, agribusiness, pharmaceutical, and biotech startup companies sped up research and development, assured that any living thing could now be granted the status of an invention.

In Europe the situation initially was different. A 1973 European Patent Convention barred patents on animals, plants, and the processes for producing them. Until the mid-1990s, the European Parliament consistently sought to exempt farmers from limits on their use of patented crops and livestock and to exclude human genes and tissues from patenting. However, the European Council (an administrative oversight body representing the separate nations of the European Union) was pushing in the opposite direction, opposing almost any limits on what could be patented. Meanwhile, despite protests from activist groups, the European Patent Office began granting numerous provisional patents for inventions involving genetic engineering, including the rights to human genes. In 1997 the European Parliament—following an intensive industry-led lobbying campaign—at last approved a resolution permitting the patenting of life forms, cells, genes, tissue, and organs. However, the issue is not yet entirely resolved.

In 1988 the U.S. PTO issued its first patent on a transgenic animal—a mouse that manufactures a human protein in its milk. Four years later, having adopted an extremely liberal policy following the Chakrabarty case, the PTO approved a broad patent to the biotech company Agracetus covering all forms of transgenic cotton. One report noted that it was as if Henry Ford had been given a patent for all automobiles.[6]

The PTO's and the judiciary's recent tendency toward extreme leniency in construing patent rights has extended to the patenting of body parts that have not been altered in any way. For example, the American firm Systemix, Inc. of Palo Alto, California, has been awarded a patent covering all human bone marrow stem cells. In 1995 Patent No. 5,397,676 was issued to the

National Institutes of Health (NIH) for the genetic material of a man from the highlands of Papua New Guinea. His tribe, the Hagahai, number only about 260 people; the NIH claimed ownership of an unmodified cell line containing the man's DNA, together with several methods for using it to detect human T-lymphotrophic virus (HTLV-1)-related retroviruses. (In 1996 the patent was withdrawn under protest.) More recently, the NIH has sought patents on human genes in about twenty other countries, usually with no provision to pay the original owners of the cells from which the DNA was extracted.[7]

As of this writing, biotech "products" that have been patented in one country or another include:

- genetically altered microbes
- techniques for genetic manipulation
- cell lines (genetically distinct cells isolated and cultured in a laboratory environment)—including those not genetically altered in any way
- plasmids, vectors, and other DNA fragments
- proteins prepared by a genetic engineering process, if they have altered properties not found in previously known proteins
- plant, animal, and human genes

The moral dilemma presented by the patenting of unmodified cell lines is highlighted by the curious story of John Moore and his patented spleen cells. Moore, an Alaskan businessman, had been diagnosed with a rare cancer and underwent treatment at the University of California at Los Angeles (UCLA). One of his physicians discovered that Moore's spleen tissue produced a blood protein that facilitates the growth of white blood cells that are valuable anticancer agents. The university created a cell line from Moore's spleen tissue and, in 1984, obtained a patent on it. Meanwhile, Moore survived his illness and sued the University of California, claiming a property right over his own tissue (by this time, the commercial value of the cell line was being estimated in the hundreds of millions of dollars). In 1990 the California Supreme Court ruled against Moore, holding that the cell line was justifiably the property of UCLA—but that by failing to inform Moore of the commercial potential of his tissue the "inventors" breached fiduciary responsibility and were liable for monetary damages.[8]

Historically, the U.S. has tended to allow a broader range of patents than other countries. But the international agreement on Trade-Related Intellectual Property Rights (TRIPS), which was part of the Uruguay Round of the General Agreement on Tariffs and Trade (GATT), extended patent rights to genetic material worldwide. This means, for example, that indigenous farmers can lose rights to their traditional seed stocks if those seeds are patented by a corporation headquartered halfway around the globe.

It also means that segments of the human genome are available for patenting. The current rush to lay claim to human genes and cell lines around the world is typified by Sequana Therapeutics' patenting of a gene related to asthma—discovered by "prospecting" among the three hundred inhabitants of a remote island in the Atlantic who have a disproportionate tendency to develop the disease;[9] and by Myriad Genetic's attempt to patent a gene that appears to cause breast cancer in some women who have a history of breast cancer in their families.[10] But these are only two examples out of scores. The patenting of human genes may serve as an incentive for investment in research, but it underscores a serious moral question: Are there ethical limits to what humans should be able to privatize, buy, and sell?

Biotech critic Jeremy Rifkin draws a comparison between the patenting of seeds and human genes and the enclosure of common lands that occurred in Europe in the seventeenth and eighteenth centuries. At that time, the village life of European peasants was overturned when landlords placed formerly public lands under private control, building fences and other barriers to the free passage of people and animals. Land became a commodity, and hundreds of thousands of peasants—now unable to grow subsistence crops on common lands—were forced to move to cities and sell their labor for daily wages. Enclosure caused people to view the world less in terms of reciprocal, traditional obligations and more in terms of profit and exploitability; it was a fundamental turning point in history, the beginning of the modern age. More recently, much of the oceanic commons and the electromagnetic spectrum have similarly been enclosed and privatized, with coastal fisheries and broadcast frequencies being leased to private companies. Now, says Rifkin, corporations are engaged in the enclosure of the genetic commons of the planet, including the human genome.

According to Rifkin, "The debate over life patents is one of the most important issues ever to face the human family. Life patents strike at the core of our beliefs about the very nature of life and whether it is conceived of as having intrinsic or merely utility value." The outcome of that debate, says Rifkin, ". . . is likely to be as important to the next era in history as the debates over usury, slavery, and involuntary indenture have been to the era just passing."[11]

Monsanto: Betting the Farm on Biotech

The manner in which "pure" science is being driven by commercial priorities in the evolution of biotechnology is perhaps best illustrated by examining how a single company—Monsanto Corporation—has become a major player in the world market for genetically modified products.

Founded in 1901 and based in St. Louis, Missouri, Monsanto is today a $9 billion-per-year business with a corporate presence in 130 countries. From its inception, Monsanto specialized in developing new chemicals and materials, including nylon and acrylic fibers, water treatment chemicals, fire retardants, coatings and adhesives, and aviation hydraulic fluids. During the Vietnam War era, the company developed and produced Agent Orange, as well as a substantial portion of the world's polychlorinated biphenyls (PCBs)—a group of chemicals eventually found to be so hazardous that the U.S. Congress banned their production in 1976. Monsanto researchers later developed Roundup, the world's best-selling herbicide (which today accounts for 15 percent of the company's sales and 40 percent of its operating profit), and NutraSweet, used in thousands of food and beverage products worldwide for "better taste and fewer calories" (with about $800 million in yearly sales).

The downside to Monsanto's success in commercial chemistry was that the company acquired a reputation as one of the worst U.S. industrial polluters. In 1992 Monsanto's products and manufacturing processes were responsible for about 5 percent of the 5.7 billion pounds of toxic chemicals pumped into the U.S. environment. With public opposition to pollution mounting and government regulations tightening, company executives saw the writing on the wall, and in the early 1990s Monsanto decided to go green. "Sustainability" and "a healthy environment" became corporate slogans. At the heart of this new strategy lay biotechnology.

In 1996 Monsanto spun off its industrial chemicals division, leaving its crop protection division as the core of a streamlined, retargeted enterprise. The company's new mission, according to Karl Sestak, assistant to Monsanto's Crop Division president, was ". . . to improve the world's capacity to produce high quality foods."[12] With a global human population approaching six billion and burgeoning environmental problems resulting from industrial farming practices, humanity needed to find new ways to feed itself in the coming century. Monsanto planned to lead the way with genetically tailored crops and environmentally friendly pesticides and herbicides. "Sustainable agriculture," its spokesmen claimed, "is only possible with biotechnology and imaginative chemistry."[13]

Monsanto began purchasing small biotech firms—Calgene, Agracetus, Biopol Business; it also bought seed companies—Cargill Hybrid Seeds (international division), Delta and

Pine Land Company (DeltaPine), Holden Foundation Seeds, DeKalb Genetics, Asgrow, and Stoneville Pedigreed. According to Monsanto's Robert Farley, the object of all these acquisitions was "not just a consolidation of seed companies . . . [but] a consolidation of the entire food chain."[14]

However, this ambitious program soon encountered a few potholes. Monsanto's first biotech product on the market (in 1993) was a recombinant version of bovine growth hormone (rBGH, trademarked as Posilac), intended to increase milk production. Monsanto literature proclaimed that rBGH ". . . helps dairy cows produce milk more efficiently, without any loss in quality or natural wholesomeness."[15]

Critics first noted that there was no need for such a product: America was already glutted with milk and milk products, to the point that the federal government was paying dairy farmers not to produce more. They also pointed out that the U.S. Food and Drug Administration (FDA) regulators who approved Posilac were former Monsanto employees who then quit their government jobs and went back to work for Monsanto.[16] Studies showed that cows treated with rBGH led shorter lives and had a greater tendency to develop mastitis (requiring the increased use of antibiotics, residues of which are passed through to the milk). Later studies also showed that rBGH-treated cows produced milk with elevated levels of the hormone insulin-like growth factor-1 (IGF-1), which has been associated with increased cancer rates in laboratory animals and humans.

Public-interest groups demanded that milk from rBGH-treated cows be labeled as such, but the FDA, in a move that stunned pure-food activists, essentially banned "rBGH-free" labeling. The rule was written by Michael Taylor, an attorney who worked for Monsanto both before and after his tenure as an FDA official. Monsanto proceeded to bring lawsuits against two dairies that labeled their milk "rBGH-free." The legal battle was costly to the dairies, and simultaneously Monsanto sent letters to dairy organizations and to retail stores threatening legal action if they dared to label products "rGBH-free."[17]

Some Florida dairy herds grew sick shortly after starting rBGH treatment. One Florida farmer, Charles Knight, lost 75 percent of his herd and claimed that Monsanto, and Monsanto-funded researchers at the University of Florida, had withheld from him the information that other dairy herds were experiencing similar problems.

Knight spoke on camera to two award-winning television reporters from Fox network affiliate WTVT in Tampa, Steve Wilson and Jane Akre, who had been hired by the station to produce a series on rBGH. The series included interviews with scientists, farmers, Monsanto spokespeople, and government officials. But before the report could air, Monsanto lawyers sent two letters to the station suggesting that Monsanto would suffer "enormous damage" if the series ran and warning of "dire consequences" for Fox if it were not canceled. Fox lawyers tried to water down the series, then offered to pay the reporters to leave their jobs and keep quiet. Wilson and Akre refused, and on April 2, 1998, they filed a lawsuit against WTVT claiming that the station violated its broadcast license by demanding that they include known falsehoods (exonerating Monsanto and rBGH) in their story.

Even before the WTVT incident, Posilac had become a public-relations nightmare for Monsanto. Due to farmer and consumer opposition, only 4 percent of America's dairy cows were being injected with the hormone. Wall Street analysts told *Business Week* magazine in 1996 that the product was a commercial failure and that it should be taken off the market.[18] As of this writing, Monsanto continues to promote Posilac.

Monsanto's next biotech products consisted of two bioengineered cotton varieties—NuCotn, containing the patented Bollgard gene, and Roundup Ready cotton, engineered to withstand multiple direct applications of Monsanto's flagship herbicide.

NuCotn plants contain a gene from the soil microbe *Bacillus thuringiensis* (Bt) that produces proteins poisonous to the bollworm. Organic growers had been using Bt for years as a natural pest control agent. Monsanto told farmers they would not have to use chemical pesticides on NuCotn, since the plants themselves would ward off bollworms. Farmers planting NuCotn paid a $32-per-acre "technology fee" over and above the price of the seed. In 1996 farmers planted nearly 1.8 million acres with the patented seeds (about 13 percent of total U.S. cotton acreage); Monsanto collected more than $50 million in technology fees.

During that year, the South experienced a hot, dry growing season. These conditions caused the cotton bollworm to flourish but also caused bioengineered cotton plants to synthesize less of the Bt protein than predicted. This combination spelled disaster for nearly half of the acreage planted with NuCotn. As a solution, Monsanto suggested spraying with conventional pesticides; this angered farmers, who had paid a premium for the hi-tech seeds.

The first commercial use of Roundup Ready cotton also yielded disappointing results. In 1997 farmers across the U.S. Cotton Belt planted the variety on more than eight hundred thousand acres. Farmers buying Roundup Ready seed had to sign an agreement promising to use only Roundup on their crop (under penalty of paying one hundred times the cost of the seed) and giving Monsanto the right to send agents onto their farms at any time to verify compliance. But cotton bolls became misshapen after the second and final Roundup application on many crops; a large percentage simply fell off the plants, making them unharvestable. At least twenty thousand acres were lost in Mississippi alone, and Monsanto had to compensate farmers for their losses.[19]

There is some indication that initial problems with NuCotn and Roundup Ready cotton can eventually be solved by fine tuning their patented plant genes. But critics of agricultural biotech have raised concerns that go far beyond the question of whether these products work as advertised. There are indications that transgenic crops can breed with local weeds, resulting in "genetic pollution" (for example, the rape-seed plant easily cross-pollinates with wild radish); at the very least, gene pollution could result in herbicide-resistant weeds. And Bollgard cotton seems destined to promote the evolution of Bt-resistant insects, thus rendering Bt useless to organic farmers. The solutions proposed so far include spraying herbicide-resistant weeds with a chemical different from the one to which they've developed a resistance and avoiding insect Bt resistance by preserving refuges of plants that lack the Bt gene. However, these strategies imply the increased use of pesticides as time goes on (which flies in the face of Monsanto's new environmentally friendly corporate image), as well as the maintenance by farmers of fields infested with non-immune pests. Such tactics are clearly problematic. [20]

Monsanto's setbacks with rBGH, NuCotn, and Roundup Ready cotton—in addition to the commercial failure of the Flavr Savr tomato, the first designer-gene vegetable to come to market, developed by Calgene before that firm was purchased by Monsanto in May 1997—must worry company executives. Nevertheless, the latter are moving full speed ahead with the development of many other new products.

The most controversial of these is the so-called "Terminator" gene technology—developed by Monsanto subsidiary DeltaPine in collaboration with the USDA—which the company expects to market by 2004. This is a "technology protection system" that causes seeds to work for only one growing season, thus preventing farmers from harvesting seeds from one year's crop to plant in the next growing season. Melvin J. Oliver, USDA molecular biologist and the primary inventor of Terminator, says, "My main interest is protection of American technology. Our mission is to protect U.S. agriculture and to make us competitive in the face of foreign competition. Without this, there is no way of protecting [patented seed]."[21]

Seed saving is an ancient and widespread practice (80 to 90 percent of wheat farmers in South Dakota rely on it, as well as virtually all traditional subsistence farmers in developing nations around the globe), but it represents a barrier and a nuisance to seed companies—particularly ones marketing patented gene-engineered varieties. Monsanto already requires that buyers of Roundup Ready seeds agree to use them only once and hires Pinkerton investigators to root out violators (seed-saving farmers in Kentucky, Iowa, and Illinois have already been forced to pay fines of up to $35,000 each to Monsanto). Seeds with the new Terminator technology will obviate the need for patent enforcement: farmers will have to buy new seed each year. This will mean increased profits to seed companies—which will pay DeltaPine for a license to use the Terminator—and far greater dependence by farmers on the commercial seed market.

However, critics see this dependence as a potential threat to world food security. Lawrence Busch, sociology professor at Michigan State University, notes that wars, civil disturbances, and natural catastrophes have historically proven capable of disrupting distribution systems or even wiping out seed supplies. "If farmers can't plant the stuff they harvest," says Busch, "and become totally dependent on [the seed companies for new seeds each year], you are really raising the ante on the possibility of mass starvation."[22]

In the view of an organic farmer I interviewed, David Letourneau of Occidental, California, "Terminator is the ultimate control mechanism. Farmers have been saving seeds for life; these guys are killing seeds for power. It's as simple as that."[23]

Monsanto and its subsidiary seed companies have plenty of incentive to develop a product like Terminator as they bring more and more patented seeds to market. Products in development include Roundup Ready corn, oilseed rape, sugar beets, wheat, and rice; disease-protected corn and wheat; cotton with genes that produce colors, thus eliminating the need for dyeing; virus-protected tomatoes; insect-protected corn and cotton; several genetically engineered vegetable oils; and improved-fiber cotton.

However, opposition to Terminator is widespread and growing. In August 1998, India's agriculture minister Som Pal told the Indian parliament that he had banned the importing of seeds containing the Terminator gene because of potential harm to Indian agriculture. And in October, the Consultative Group on International Agricultural Research (CGIAR), the world's largest international agricultural research network, banned Terminator technology from all its crop-breeding programs, citing "(a) concerns over potential risks of its inadvertent or unintended spread through pollen; (b) the possibilities of sale or exchange of unviable seed for planting; (c) the importance of farm-saved seed, particularly to resource-poor farmers; (d) potential negative impacts on genetic diversity; and (e) the importance of farmer selection and breeding for sustainable agriculture."[24]

On October 27, 1998, Monsanto CEO Robert Shapiro addressed the "State of the World Forum" in San Francisco, touting his company's genetic engineering programs as the solution to problems of world hunger and pesticide pollution. Following his speech he was confronted by antibiotech protesters who threw a vegan tofu-creme pie in his face.

This embarrassment capped a year of bad news for the company. A widely publicized merger with pharmaceutical giant

American Home Products had fizzled. Monsanto was suffering growing public-relations and marketing problems in Europe and South America—especially relating to the Terminator. The company was now heavily in debt and its stock value had plummeted 30 percent in the preceding weeks. And financial analysts predicted that it was ripe for unfriendly takeover by one of the other large "life science" transnationals (Dow, DuPont, or Novartis)—largely because of overinvestment in genetic engineering.

Leave It to the Market?

The example of Monsanto offers us a window into the world of corporate agricultural biotech. The company is one of the major players in the industry but not the dominant one. As of 1997, its competitors include Pioneer Hi-Bred, Mycogen Seeds, Ciba Seeds, DowElanco, and AgrEvo. Together, these six agrochemical corporations accounted for nearly 40 percent of world seed sales. Each year, smaller companies are bought up or go out of business. In 1998, for example, AgrEvo purchased Cargill Hybrid Seeds North America for $650 million; and Dow Chemical purchased controlling shares of Mycogen Seeds (which then swallowed up two Brazilian companies, Hibridos Colorado and FT Biogenetica de Milho). Early this year (1999), Dow bought Pioneer Hi-Bred as well.

Like Monsanto, the other big biotech seed companies claim that their primary objective is to feed the world's growing population. However, there are surely other motives at work as well—including the promotion of pesticides manufactured by the same or related companies. A survey by the Union of Concerned Scientists found that 93 percent of genetically engineered food crops then undergoing field tests were intended to make food growing and processing more profitable, while only 7 percent focused on nutritional or flavor improvement.[25]

Regardless of actual motives, corporate agricultural biotech policies are bound to affect both consumers and farmers. Already about 60 percent of all foods sold in the U.S. are genetically engineered or contain a genetically engineered ingredient. It is estimated that by the year 2005, over 80 percent of all foods grown in the U.S. will be genetically modified. While biotech companies insist that their gene-engineered varieties are novel enough to deserve patents, their lawyers have simultaneously convinced the FDA that such foods are "substantially equivalent" to non-engineered counterparts and therefore do not require labeling or safety tests. We consumers thus become the guinea pigs in a vast science experiment involving nearly the entire human food system.

Organic farmer David Letourneau told me he believes that "the single biggest moral issue here is labeling. Unless genetically engineered foods are clearly labeled as such, there is no freedom of choice. Now, I personally believe that there are so many ecological problems with agricultural biotechnology that it should just be outlawed. But consumers should at the very least be given knowledge of what they're purchasing and eating. The denial of that freedom of choice is clearly an ethical issue of huge dimensions. It prevents people from following their own dietary philosophies and taking responsibility for their own health. If you have a fish gene in a strawberry, how can you be a vegetarian?"

The biotech industry actively opposes the labeling of genetically engineered foods, even though polls consistently show an overwhelming majority of consumers support it (a poll by the biotech corporation Novartis in February 1997 revealed that a full 93 percent of Americans want all genetically engineered foods labeled as such). The executives' reasoning is clear: if bioengineered foods came with a special logo or warning label, consumers would be likely to avoid them. As the head of Asgrow seed company (a Monsanto subsidiary) candidly admitted to the press recently, "Labeling is the key issue. If you put a label on genetically engineered food you might as well put a skull and crossbones on it."[26]

Consumers can avoid most genetically engineered foods by buying certified organic products. However, this is no sure solution. Letourneau, who is also the Chairman of the Government Affairs Committee of California Certified Organic Farmers (CCOF), the state's organic certification agency, cautioned: "You have to take into account the fact that food that's labeled 'organic' is allowed to have some ingredients that don't meet organic standards. So your 'organic' granola or corn chips could well be made with genetically engineered vegetable oils. Without strict labeling laws, nobody's going to be able to tell what they're eating."

Not only consumers but farmers as well are feeling the brunt of change as biotech proliferates. Only decades ago, farming required detailed, intergenerational knowledge of soil, weather, and traditional seed stocks. Gradually, with the development of industrial agriculture, farmers have come to rely less on their communal knowledge base and more on advice from experts in universities, agrichemical companies, and government agencies. Now, in order to plant new gene-altered crops, farmers must sign agreements to grow strictly according to instructions, using designated pesticides and herbicides, saving no seeds. Farmers have little incentive to retain traditional knowledge or heirloom seed varieties—the products of centuries or millennia of observation and experiment. In the process, they are becoming merely a pool of deskilled labor. As Verlyn Klinkenborn remarked in a *New York Times* Editorial Observer column, "[This strategy] may make sense in terms of corporate profits, but it makes no sense at all in terms of the resources that really matter to the health of the land and the people who live upon it."[27]

Letourneau expresses his own views of agricultural biotech with characteristic bluntness: "Farmers are going to be bio-serfs and the entire food production process will be controlled by giant corporations. The corporations want to control our food supply. I'm not exaggerating. Otherwise they wouldn't be patenting seeds; they wouldn't have developed the Terminator seed. They're reducing life to a commodity, and that's scary."

Monsanto and its competitors are primarily interested in agricultural biotechnology, but similar commercial pressures are observable in the field of medical biotech, where companies and research centers are quickly buying up rights to potentially profitable bits of the human genome. Many medical biotech patent applications lay claim to the genes that appear to trigger certain diseases, so that profits can be made on tests for those genes.

Thus Myriad Genetics hopes to patent a "brain cancer gene." Meanwhile, Darwin Molecular has already patented a "premature aging gene"; Millennium Pharmaceuticals has a patent on an "obesity gene"; Human Genome Sciences has patented an "osteoporosis gene"; Axis Pharmaceuticals has patented a "blindness gene"; and Duke University has patented an "Alzheimer's gene."

Only the most extreme critics of biotechnology would oppose genetic research aimed at eliminating diseases like Alzheimer's or cancer. However, the search for profits may be leading medical biotech researchers in competing pharmaceutical companies to patent as many genes as possible before they even know the meaning or potential uses of these genes. Researchers have become genetic prospectors, searching for gold in the mine of the human genome.

Corporations and Bioethics

Biotechnology promises to change us in many ways—what we eat, how we produce food and basic materials, how we treat our environment, how we think about disease and health, how we deal with criminality and insanity, how we conceive of ethics and spirituality, and perhaps even what we look like and how we reproduce. During the course of this immense transformation of society, who will win, who will lose—and who will decide? Should we leave such monumental decisions to the marketplace and to major corporations? Should private companies be making huge profits by monopolizing human genes or patenting entire living species?

The various applications of biotechnology will inevitably have unintended consequences, and some are likely to be catastrophic. Who will take responsibility for unforeseen ecological, social, or medical problems? Typically, those who introduce new technologies profit the most from them, while society as a whole bears the later costs.

We live in an age dominated by corporate power—a fact which, by itself, raises plenty of ethical questions. Corporations directly or indirectly wield control over public information media, elected officials, regulatory agencies, and universities. Corporations are not living entities, yet courts have given them the legal and political rights of mortal persons (e.g., free speech). They can operate in many countries at once. The yearly incomes of many corporations are now greater than the gross domestic product of many nations. Yet corporations are not democratic institutions; while their actions may affect the public at large, the public has little or no say in their operation. Corporations are created in part to shield stockholders and executives from responsibility. In many states (such as California), an individual can be imprisoned for life for committing three felony crimes; many corporations commit dozens or hundreds of felonies and merely pay insignificant fines. Thus scores of social theorists from across the political spectrum warn that the growth of corporate power threatens the very foundations of democracy in America and around the world.[28]

Corporate apologists argue in return that the marketplace is amoral and leads society in no particular direction. Furthermore, it is thoroughly democratic, because consumers cast their "vote" by the ways in which they spend their dollars. However, the food-labeling issue undermines this position. If the market is "economic democracy" in action, then lack of adequate food labeling is equivalent to presenting voters with ballots that omit candidates' names or party affiliations. It is in the economic interest of corporations to prevent the labeling of genetically altered foods, and their overwhelming economic and hence political power has so far enabled them to do so. It seems that in this instance, the marketplace—that is, the power of money—has blocked and undermined democracy, rather than promoted it.

Today's corporations have gained control, via patent rights, over the genetic basis of the human food supply (which has consisted until recently of thousands of traditional varieties developed over the millennia by farmers around the globe) and over much of the human genome. Even if genetic engineering is not morally objectionable for other reasons, this concentration of the ownership of fundamental genetic resources in fewer and fewer hands is deeply troubling.

CHAPTER FIVE

Exploring Our Intuitive Responses to Biotech: "Yuk" or "Wow"?

THE IDEA OF engineering genes in plants, animals, and humans evokes feelings that are sometimes hard to articulate or to evaluate objectively. On one hand, when contemplating some of the possibilities (a cow with extra udders or a pig fitted with human genes so that its organs can be "harvested" for medical transplants), we may experience what might be called a "yuk" response—a vague sense of something being horribly wrong or out of place, or of something sacred being violated. On the other hand, the idea of being able to alter organisms in basic ways (e.g., to engineer bacteria to produce useful biochemicals) may evoke a breathless "wow"—an exhilarating appreciation for humanity's godlike power.

Yet when we recoil with an instinctive "yuk," just what is it that we sense is being violated? And how is the "godlike power" of a biotechnician any more mysterious or enthralling than that of a nuclear physicist, a billionaire investor, or the president of a country? Questions like these challenge any simplistic approach to the moral evaluation of biotech.

It may seem silly at first to mention gut responses in the context of evaluating something as complex and important as genetic engineering. My own first instinct is to rationally take stock of the benefits and dangers of biotech, weigh the evidence according to objective criteria, and arrive at justifiable conclusions.

Certainly our judgments about genetic technology need to be informed by acquaintance with scientific facts and theories. Yet in the end perhaps our choices will still issue from that part of ourselves that transcends rationality. And maybe this is as it should be.

If we want to understand biotech from a spiritual and moral perspective, we must come to terms with our intuitive feelings, because these are pathways to the very essence of morality and spirituality. Religion thrives on feelings of awe, dread, guilt, and humility. And no matter how avidly we may try to develop moral principles by applying logic and reason, the exercise inevitably leads us back to these same feelings, which register more in the pits of our stomachs than in our detached minds. The ability to feel compassion for the suffering of another person or animal, to feel awe when contemplating the natural world and the cosmos, or to feel a sense of injustice or outrage when we see something whole and pure being needlessly violated—these are the soul's equivalent of sight and hearing, and the practice of spirituality is about cultivating and refining these senses.

Granted, even if the moral or religious sense is every bit as real as the senses of hearing and sight, it is also just as easy to trick or confuse. The citizens of Salem, Massachusetts, who conducted their infamous witch hunts in the seventeenth century were no doubt acting upon strong feelings—feelings of terror and loathing which they trained upon innocent scapegoats, with horrific consequences. It's no wonder, then, that we sometimes mistrust our moral or religious sensibilities. Nevertheless, to dismiss them altogether would be to reject an important means of perception; it would be as foolish as to disbelieve everything we see out of fear that we might be subject to optical illusions.

Applying our moral sensibilities to the dilemma of biotechnology is not a simple matter. The subject cannot be taken in entirely at one glance; it is complex and multifaceted. Our perception of medical biotech, for example, might be quite different from that of agricultural biotech, just as our moral take on the cloning of farm animals might differ from that on the engineering of bacteria to produce valuable biochemicals. The moral questions surrounding biotech are also multilayered: at the surface we deal with simple, discrete, black-or-white questions (is it right to perform biotech experiments on large laboratory animals, knowing that the experiment will involve the suffering or death of the animals?); but as we penetrate toward the core of the subject, we are faced with more subtle issues having to do with the nature and sanctity of life and the proper applications of human intelligence.

A Morally Neutral Tool?

The most superficial approach to the moral assessment of biotech would be to say that the moral sensibility simply does not, or should not, apply to biotechnology per se; that biotech is just a tool, and no tool is inherently either moral or immoral.

Ethics come into the picture only when we consider how the tool is used.

This view is widespread and rational. It explains the instinctive "yuk" that so many people feel when contemplating the prospects of a gene-altered future as arising merely from the fear that the tool will be misused—or from squeamishness or unfamiliarity. After all, many people had a similar "yuk" response to automobiles and airplanes in the days when those were novelties, believing that humans were "not meant to" travel so quickly or fly through the air. The thrill of newfound power—the "wow" response—has likewise accompanied the introduction of almost every new tool, from the flint-tipped spear to the microchip. Every step along the path of technological progress has aroused both fear and excitement; the fact that biotech evokes these feelings reveals only its novelty, not whether it is inherently bad or good.

Australian medical researcher G. J. V. Nossal pursues this line of thinking in his book *Reshaping Life,* stating that ". . . it is unfair to blame science and technology for ills in the human condition that are as old as mankind: for undue aggression, selfishness, greed and a chronic incapacity to live up to one's highest aspirations. . . ."[1] There have always been people with an irrational distrust of science and technology, Nossal points out. While specific concerns about the abuses of certain technologies may be valid, sweeping objections to any scientific or technological advance—including the techniques of cloning and gene splicing—should be disregarded. The American Council on Science and Health makes the same argument in its pamphlet *Bio-*

technology: An Introduction, saying that "Biotechnology techniques by themselves are neither good nor evil—but they can be used toward either end."

However, the idea that all technologies are morally neutral—a view that especially appeals to many people who pride themselves on their reasoning powers—is questionable on purely logical grounds. Technology theorist Jerry Mander frames the matter this way:

> If you once accept the principle of an army—a collection of military technologies and people to run them—all gathered together for the purpose of fighting, overpowering, killing, and winning, then it is obvious that the supervisors of armies will be the sort of people who desire to fight, overpower, kill and win and also are good at these assignments: generals. The fact of generals, then, is predictable by the creation of armies. The kinds of generals are also predetermined. Humanistic, loving, pacifistic generals, though they may exist from time to time, are extremely rare in armies. It is useless to advocate that we have more of them.
>
> If you accept the existence of automobiles, you also accept the existence of roads laid upon the landscape, oil to run the cars, and huge institutions to find the oil, pump it and distribute it. In addition you accept a sped-up style of life. . . .
>
> If you accept nuclear power plants, you also accept a techno-scientific-industrial-military elite. Without these

> people in charge, you could not have nuclear power. . . .
>
> If you accept mass production, you accept that a small number of people will supervise the daily existence of a much larger number of people. You accept that human beings will spend long hours, every day, engaged in repetitive work, while suppressing any desires for experience or activity beyond this work. . . .
>
> If you accept the existence of advertising, you accept a system designed to persuade and to dominate minds by interfering in people's thinking patterns. . . .
>
> In all of these instances, the basic form of the institution and the technology determines its interaction with the world, the way it will be used, the kind of people who use it, and to what ends.[2]

Societies tend to reshape themselves to accommodate the tools they adopt. As historian of technology James Burke and science writer Robert Ornstein put it in their book *The Axemaker's Gift,* every new tool "first changes the environment and then changes our way of thinking and our values. . . ."[3] Thus a society supplied with electrical power from a few huge nuclear reactors is more likely to be governed by a centralized bureaucracy than one that derives energy from thousands of small windmills or solar panels. Mander goes so far as to say that the centralized nuke-powered society is likely to be controlled by a "techno-scientific-industrial-military elite" class. Experience bears this out: while some countries that have nuclear reactors (including the U.S.) are titular democracies, in every instance there is continual political tension between an elite that promotes, regulates, and operates nuclear power plants and grassroots citizens groups that wish to decentralize energy production and replace nuclear power with more environmentally benign solar and wind power. The proponents of nuclear power often use authoritarian tactics (expensive propaganda campaigns and secret governmental policy meetings), while their opponents prefer more democratic tactics (demonstrations and public hearings). Many people would agree that the choice between a decentralized and democratic society on one hand and a centralized, authoritarian society on the other is at least partly a moral one. Therefore, in this case, it is clear that our society's adoption of a certain tool (the nuclear reactor) has had ramifications that are distinctly moral in character—even if it's not a simple matter of "solar good, nuke bad."

With regard to biotech as well, we must assume that the technology will *by its very nature* imply a range of social consequences as we adopt it. We would be wise, then, to inquire what those consequences might be and to think about whether they are consequences with which we want our children and grandchildren to live.

As we have seen, biotech develops from and reinforces a certain way of looking at the world—a mechanistic, reductionist, utilitarian view not only of genes but of life itself. This is part of the source of both our "yuk" and "wow" responses to it. For many people, the act of treating an animal such as a cow or a sheep as merely a living machine to be genetically manipulated violates a sense of organic wholeness and of compassionate connection with another conscious being. Yet this utilitarian attitude also gives us power over nature in ways previously undreamed of—as in engineering cows or sheep to produce pharmaceuticals in their milk. If we adopt biotechnology, we thereby ensure that future generations will tend toward a more utilitarian view of nature and have less of an inclination to regard other organisms with wonder and compassion.

In addition to embodying and reproducing a particular world view, biotechnology is likely to cause a society using it to mutate in other predictable ways. People who live in a biotech society will be more likely than ourselves or our ancestors to treat their own bodies as products to be redesigned and improved upon. Wealthy people will be in position to take greater advantage of the ability to "improve" their offspring than will poor people, and so class divisions will likely become genetic as well as economic.

The question of whether we want to move in these directions, and *how far* we wish to go, is a moral—perhaps even a spiritual—question. However, merely acknowledging the question as having a moral and spiritual basis does not suggest an answer. Will we listen more to our inner "yuk" or our inner "wow"?

Taking the Moral Pulse

Biotech may have individual applications that we judge to be beneficial or harmful. However, our consideration of its moral impact should not be limited to those specific applications, because the technology per se will inevitably reshape society and human consciousness. It is possible that biotech itself may be both good and bad in its effects, but it will *not* be neutral. Indeed, to treat biotech *as though it were* morally neutral may be merely a mental diversion, a means of avoiding responsibility. If we are to act as morally and spiritually responsible agents, we should evaluate and debate the moral impact of biotechnology as early in its development as possible; the longer we wait, the less control we will have over it. And this process of evaluation must take into account not only scientific predictions about likely benefits and harms, but also our own spontaneous, intuitive feelings.

Immediately, we're in difficult territory. Assuming we are interested in seriously considering people's moral and spiritual feelings about biotech, where should we start? With a poll of all citizens? With a discussion among interested parties (industry representatives, regulators, activists)? Or with an inquiry by moral specialists (ethicists, clergy, and philosophers)?

Polls of the citizenry consistently show uneasiness with biotech in most of its applications. As we've already seen, a recent Novartis poll showed an overwhelming majority (up to 93 percent) favoring the labeling of all genetically engineered foods. A somewhat smaller majority would avoid all such foods if so labeled. In a USDA-funded poll released in January of 1996 on recombinant bovine growth hormone (rBGH), 94 percent of consumers said rBGH-derived dairy products should be labeled and 74 percent believed that the drug was unsafe. Meanwhile, according to a February 1998 *Time*/CNN poll of 1,005 adults, 93 percent of Americans oppose the cloning of humans and 66 percent oppose the cloning of animals.[4]

Some may protest that this popular negative perception of biotech arises largely from ignorance. Should science proceed by majority rule? Are all voices equally worth hearing? Some would answer no, that those who know the most about the technology should have more decision-making authority over it—and in this instance those who know the most are primarily scientists and corporate managers. This approach would virtually guarantee a context favorable to unhindered implementation. However, if the people who know most about the technology and have the most authority over it are also the people who have the most to gain from it, then an even-handed moral assessment is virtually excluded from the start. Thus it is important that the technology be regulated by some independent yet knowledgeable body or bodies (e.g., government agencies such as the USDA and the FDA). And every effort should be made to educate the general public about the technology and its implications.

Unfortunately, the corporations that have the most to gain from biotech have also established dominance in the arenas of government regulation and public education. Corporations control many university appointments and fund most relevant symposiums and conferences. Personnel cycle back and forth between industry and government regulatory agencies. Agricultural biotech companies send public relations representatives to "educate" young people in high school and college classes throughout the country (when my wife was working toward her horticulture degree at a local college, her class was visited by a sincere, knowledgeable young woman from Monsanto, who offered an hour-long paean to the benefits of gene-enhanced farming). And medical schools receive regular visits from pharmaceutical biotech reps.

Most people would agree that moral questions should not be decided on the basis of whose position is backed by the most wealth and political leverage. Often the most compelling moral voice issues from members of society who are relatively powerless. This was driven home to me in my conversation with David Letourneau, who told me that his own views on biotech were shaped largely by conversations with mothers. "When the controversy first arose, one guy from the Organic Trade Association said, 'We don't want to say yes to genetically engineered food and we don't want to say no.' The people who were against it were the mothers—Mothers for Natural Law. They were adamantly opposed. To which I said, 'I'm with you.' These are the mothers of our children, and as far as I'm concerned they are the ultimate arbiters of our standards. I'll believe them before I'll believe the representative of some manufacturing company that's mostly concerned about the bottom line."[5]

One can envision panels of mothers empowered to decide the fate of thousands of cloning and gene-splicing experiments. That might in fact be an excellent social mechanism for finding solutions to the ethical problems posed by biotech and perhaps nuclear energy, pollution control, and military policy as well. However, one can also envision the objections from industry: Shouldn't other interested parties have some say in decisions? Why mothers? Why not uncles or cousins?

Even most scientists and corporate managers would likely agree that the discussion needs referees. But who can be trusted by *all* parties to be knowledgeable and impartial?

The referees that society has chosen for the time being are professional ethicists—academic specialists (mostly philosophers) who make it their business to debate what's right and wrong and why. Over the past two decades, the emergence of biotechnology, with its attendant moral quandaries, has triggered the concurrent development of the discipline of bioethics. Today several universities maintain departments or programs in bioethics, the most prominent of which is the Center for Bioethics at the University of Pennsylvania.

Even a cursory search of the robust bioethics literature shows that most attention is devoted to medical issues. This focus is apparent in the writings of Arthur Caplan, Director of the Center for Bioethics. Caplan favors genetic research in general and the Human Genome Project in particular; and though he opposes some uses of biotech (human cloning and cosmetic germ-line gene therapy), he believes that the dangers of the genetic technologies have been overstated by critics. He writes:

> A few of those who doubt that humankind knows what to do with more information about its own hereditary makeup or who simply believe that it is unnatural to mess around with genes sometimes try to arouse legislative or public concern by spinning scenarios in which man-animal chimeras slink out of the corridors of MIT, Cal Tech, Genentech, or Fort Dietrich to commit maniacal man-animal misdeeds against hapless humans. If such grim scenarios aren't scary enough, the occasional critic resorts to even more horrifying futuristic timeworms in which

> hordes of clones derived from the embryos of businessmen, sports stars, and politicians (no attempt is made to mitigate the horror) descend on an unsuspecting and defenseless world. In the most hyperventilating form of such criticism warnings are issued that if the genome project is not stopped now the result will inevitably be a planet teeming with millions of knockoff copies of Adolph Hitler, Genghis Khan, Saddam Hussein, Idi Amin, and Joseph Stalin.[6]

Caplan coolheadedly advises: "Do not fall for all the hype. Do not let those who learned about cloning from Woody Allen, Gregory Peck, and Michael Keaton frighten you into thinking that science and technology must inevitably be our master. Human beings can control the technologies that they invent."[7] The way to do so is by paying attention to specific ethical problems, rather than overgeneralizing. With regard to cloning, for example, Caplan continues:

> If we really think that it is an offense to human dignity to have people bred by design, if it seems nuts to let anyone create a human being just to have a place to get spare parts, or if we really do not want grieving parents trying to 'restore' a lost child by making a physical copy of the child's body, we really can bring such activities to a grinding halt. How? Slap stiff penalties on human cloning, let researchers know that experiments on human cloning will not be published and pull all Federal and foundation sup-

port for human cloning research. For all practical purposes human cloning will grind to a halt."[8]

When it comes to human germ-line gene therapy—altering the genetic code in a human gamete or fertilized ovum so that all of the cells of the individual that develops from them will harbor the change—we should give some attention to the problem of defining "disease." Clearly (in Caplan's view) it would be wrong to withhold a genetic cure for fatal or crippling illnesses like Tay-Sachs, thalessemia, or Hurler's syndrome. "If it were possible to eliminate a lethal gene from the human population by germ-line alterations, is there any convincing moral reason why this should not be done?" he asks. "If those carrying a lethal gene request treatment so that they are able to reproduce without guilt or fear, ought not health care providers feel not reluctance but a duty to help them?" The danger lies in an increasing tendency to define shortness, myopia, baldness, and other common characteristics as diseases. Caplan's solution? Define *disease* better. "The way to handle legitimate concerns about the dangers and potential for abuse of new knowledge generated by the genome is to forthrightly examine what are and are not appropriate goals for those who provide services and interventions in health care. There is nothing sacrosanct about the human genome. It is only our inability to openly and clearly define what constitutes disease in the domain of genetics that makes us feel that intervention with the germ-line is playing with moral fire."[9]

In his book, *Bioethics: A Primer for Christians,* Gilbert Meilander writes that "if anything amounts to 'playing God' illicitly, germ cell therapy might seem to." Nevertheless, the technique offers obvious benefits; it treats not just one sufferer of a disease but all her descendants as well. "To draw back in fear here might seem to be the sin that was once called 'sloth'—an unwillingness to seize new possibilities and a readiness to fall back into the safe and familiar." Still, Meilander eventually arrives at a different conclusion from Caplan's: "We are not only free, self-determining beings. We are not self-creators, and some limits of our finitude ought to be respected. . . . In this case, I suggest, the most truly human and humane exercise of our freedom will be the courage that says no when asked to make humankind itself our patient."[10]

In reading Caplan, Meilander, and many other bioethicists, one gets the sense that these people have been thinking long and hard—and generally quite lucidly—about genetic issues that are likely to interest physicians and patients. But this narrow focus omits problems that may actually be of broader and more immediate concern than that of human cloning. An article posted on the Center for Bioethics Web site, "Ethical Issues in Genetics in the Next 100 Years" by Center for Bioethics Internet Project Director Glenn McGee, contains no mention whatever of the fact that most Americans are eating foods with genetically engineered ingredients *now* without knowing it and offers no consideration of the ecological effects or the health risks that statistic implies. Nor does McGee offer any discussion of the effects of biotechnology per se on our attitudes toward life, animals, plants, or our own bodies.

When I called the Center for Bioethics to inquire about this apparent oversight, a friendly staffer referred me to the Center for Agricultural Bioethics at Iowa State University, at whose Web site I found only a peripheral discussion of genetic engineering and no papers of a depth and erudition to match those of Caplan and McGee. I came away from my search with the sense that if biotechnology *does* pose significant threats to society, human health, and the ecology of the planet, few professional bioethicists will be of much help in warning us.

Spiritual leaders constitute another potential source of ethical advice regarding biotech. These individuals have likewise specialized in honing their moral and spiritual sensibilities. Since religion has shaped humanity's sense of morality in the past, it makes sense that we should look to it now for help in navigating the waters of technological change.

Yet discovering what religious leaders think of biotech turns out to be a tricky matter. Many simply haven't looked into the subject enough to develop an informed opinion. Among those who have, views often diverge. Religions that base their doctrines on scriptural texts face the problem that nowhere in the Bible, the Qur'an, the Talmud, or the Sutras does it say, "Thou shalt not splice DNA," or "It is thy destiny to clone sheep." Even among evangelical Christians—who tend to be fairly homogeneous in their opinions on most social issues—one finds opposing views on biotech: where one interpreter cites the Genesis 3:6 warning that humans should not attempt to be "as gods" as an injunction to refrain from manipulating genes, another cites humankind's scriptural responsibility to "replenish the earth and subdue it" (Genesis 1:28) as a mandate to engineer at will.

Nevertheless, a survey of opinions from informed clergy and religious scholars—conflicting though they be—is essential to developing a moral assessment of biotech.

Thou Shalt Not Tinker . . .

Some religious leaders are inclined to look at any interfering with nature as devilish. This tendency predates the arrival of modern biotechnology: when Luther Burbank claimed to be able to create new kinds of vegetables, Christian fundamentalists gathered on his front lawn to pray for him. Today, Christian opposition to tinkering with the divine plan is similarly aroused by genetic engineers' deliberate rearrangement of plant, animal, and (potentially) human DNA sequences. However, among various denominations and individual theologians, this opposition ranges in intensity from utter abhorrence to mild caution.

Although twenty-one Catholic bishops joined other religious leaders in signing a statement calling for a ban on all genetic engineering in 1983, the Catholic Church has focused its ethical attention primarily on medical and reproductive applications of biotech. In March 1998, Pope John Paul II condemned genetic engineering as potentially providing techniques that could be used by "totalitarian groups" to "violate human rights." The pope has said he is opposed to the "manipulation of life, at the service of boundless ambition, which deforms the aspirations and hopes of mankind and which only increases its suffering." While the

Vatican strongly opposes cloning, pronouncing it a serious attack on human dignity, it nevertheless regards genetic research into illnesses as possibly useful.

Donald Conroy, a Catholic theologian and founder of the North American Coalition on Religion and Ecology, told me: "Until recently, the Vatican has focused most of its critical interest in biotechnology on issues having to do with human reproduction. Now there is concern about transgressing the boundaries between species, the divine design of nature. The Catholic Church is also warning scientists that they have to be responsible for what they do: there is no such thing as pure science for science's sake; all science has social and moral implications." Conroy himself is deeply concerned about agricultural biotech: "People in religious communities have a right to know what's in their food. This is a very basic right. Moreover, we should respect God's creation. Right now, people who are using this technology, inserting genes, don't know where on the genome the new gene will end up, and they don't know its full range of effects. You may want to extend the shelf life of tomatoes with a fish gene, but there may be consequences to the web of life that we don't even imagine. Or it may have unanticipated health consequences for consumers—allergic reactions, for example."[11]

Among Protestants, views on genetic engineering are slow in coalescing. Dr. John Reigstad, a Lutheran clergyman and lecturer on religion at Wartburg College, told me: "We're just scraping the surface of these issues, because churches and religious traditions simply haven't had these issues before them previously. If you look at the history of religions, there is often a component of concern about food, yet there hasn't been an outcry from the religious communities about the genetic engineering of foods. My explanation for that is simply that people haven't been aware of the issues. What's happening may not be a secret to the scientific community, but it has been for scholars and laity in religious communities."[12]

Reigstad, who became concerned about the genetic engineering of food after reading an article in the London *Daily Telegraph* by the Prince of Wales,[13] agreed to be named as a plaintiff in a lawsuit against the FDA for its refusal to require the labeling of genetically engineered foods. "I don't pretend to be an expert on this subject," says Reigstad, "but it seems to me that there are certain features that anyone can see, even at a distance. There is an issue of power here—the political power of who is actually making the decisions in the laboratory and in the government. There is also a problem with knowledge—not so much with what we know now, but what we don't know yet—and with the lack of public awareness of what is being done, especially with the genetic engineering of food. One of the revolutions of religion was Martin Luther's translation of the Bible into a language that the people could understand. Something similar needs to happen now with science: we need to bring the information to the people, so they can begin to understand the issues and help make decisions.

"There may well be unforeseen and irreversible consequences with these technologies, yet we're moving ahead extremely quickly," says Reigstad. "I think we have to look carefully at the pace and timing and at how the whole process is being driven by economics. There's the temptation to come up with discoveries in the laboratory that will be profitable, and it seems to me that people are pushing these things too quickly from the laboratory into the marketplace."

Colin Gracey, Episcopal chaplain at Northwestern University and a board member of the Council for Responsible Genetics, is another plaintiff in the same lawsuit against the FDA. Like Reigstad, he sees the government's refusal to require the labeling of genetically altered foods as a moral issue: "We ought to be candid and trust the American people. Some things are being done in food production whose implications are not known even by the researchers, but they're going forward anyway. We should be concerned not just about what benefits biotechnology will yield; we should also focus on the questions we should be asking to test the assumptions and theories.

"In my view, the FDA is mandated to protect the people and to test food products. But in this case they are saying, 'We'll allow this whole new class of products to go to market untested, and if there are problems we'll deal with that later. Moreover, we'll trust the companies developing the products to do the testing.' I think this attitude partly results from cutbacks in funding for regulatory agencies during the 1980s. They're just not adequately staffed to look carefully at a whole new technology. But this is not what the FDA has traditionally done, and this is why I signed on as a plaintiff in a friendly lawsuit against the Agency—to encourage them to do their job."[14]

With regard to the issue of patenting life forms, Gracey told me: "The standard argument is that, if you don't patent, you're not going to get the infusion of money to do the research. Then there are those who say that, in the grand scheme of things, the life of the patent—seventeen or twenty-one years—is not that long. After that, the information is anyone's to use. But from a theological point of view, the patenting of life forms is a kind of threshold. Beyond it, we run the danger of so objectifying and commodifying life that we begin to treat it as less than what it really is, which is a miracle and a mystery."

Gracey, who is trained as a medical bioethicist, also holds skeptical views about human germ-line gene therapy. "If we proceed with genetic modification of humans and something goes wrong, are we prepared to deal with the consequences—to care for that harmed fetus or individual? Suppose the evidence of harm emerges twenty years later. What are the structures of government that would enable us to follow through on the process in a supportive way? With Dolly, there was one success among 276 failures; the failures were simply discarded. But we can't do that with people. Who's responsible here? Is it the laboratory? Society as a whole?"

Many evangelical Christians are similarly critical of biotech. In 1989 the evangelical journal *Christianity Today* carried an article that declared: "Genes are a core that should not be monkeyed with." And an essay in the Jehovah's Witness *Plain Truth* questioned whether man would crown himself "the new God." According to the article, people centuries from now "will look back at our age and shake their heads in utter amazement. . . . They will wonder how we could possibly have believed that man alone was capable of solving his problems of disease. . . . The real God, not the one fashioned by man's religion and cloned in our image . . . will give us all the good things the genetic revolution promises."[15]

The engineering of food plants with genes from insects, bacteria, animals, and humans violates the dietary rules of several religions. Vegetarian Jains, Buddhists, and Hindus object to animal genes in fruits and vegetables, while Jews and Muslims object to foods containing genes from prohibited animals. Buddhist scholar Ron Epstein quips that the insertion of human genes into vegetables (as is occurring in China) means that "you can now be a vegetarian and a cannibal at the same time!"[16]

In 1998 the Beth Shalom Synagogue of Fairfield, Iowa, issued a "Resolution to Protect the Dietary Rights of American Jews," which states:

> We are opposed to genetically engineered food on the basis of universal religious principles and also on the basis of Jewish dietary law. As a matter of general principle, we believe that the artificial transfer . . . of genetic material between species that are naturally prevented from cross-breeding is a violation of God's law and a reckless disruption of His intricate plan. . . .
>
> In addition to the general principles stated above, we feel an obligation to avoid foods containing any gene from a species prohibited by Jewish law.[17]

Biotech industry apologists argue that gene sequences are merely strings of molecules, not whole plants or animals. The basic chemical structure of DNA is the same in all living things.

Therefore a pig gene that has been inserted into a maize plant doesn't carry any special essence of "pigness" that would cause the maize to become part animal. However, religious critics reply that the act of shuffling genes between plants and animals can only be interpreted as a blurring of species boundaries that their traditions insist should be kept distinct.

Devotees of Eastern religions hold criticisms similar to those of Christians and Jews, although they often diverge in subtle ways. John Fagan, the geneticist who refused a $614,000 biotech research grant from the National Cancer Institute in 1994, is a follower of Maharishi Mahesh Yogi. In his book *Genetic Engineering: The Hazards/Vedic Engineering: The Solutions,* Fagan argues that "of the technologies now in use, genetic engineering is especially dangerous because many of the most common applications of this new technology threaten to generate unexpected, harmful side effects that cannot be reversed or corrected, but will afflict all future generations." Fagan calls for a complete moratorium on the implementation of most medical and agricultural biotechnologies and an exploration of alternatives based in "natural law"—which he associates with Maharishi's Vedic science. Fagan warns of interference with the natural course of human evolution through germ-line genetic engineering and of ecological risks from agricultural biotech. He argues that millennia-old Vedic science offers a safer, preventive approach to health and a sustainable model for agriculture.[18]

Even in the New Age community, which has been inspired by Teilhard de Chardin's vision of technology as an aspect of the spiritual evolution of humanity, many consider genetic engineering problematic. The Lucis Trust, an organization founded to spread the Aquarian teachings of Alice Bailey, publishes the *World Goodwill Newsletter,* which printed a special issue on the dangers of the new genetic technologies. Among other aspects of biotech, the unamed authors criticized the engineering of animals to produce useful chemicals or to serve as donors for organ transplants to humans:

> Because the main purpose of these . . . projects is to benefit humans, they reinforce the assumption that only human beings are of major value and that all other forms of life can be subordinated to human ends. This is an assumption which a growing number of people are questioning, proposing that every creature within the great web of life is intrinsically valuable. If we accept this premise, then every relationship which humans enter into with other creatures should be characterized by goodwill. At the very least this would call for national and international regulatory processes to govern genetic engineering experiments. . . .[19]

In a conversation at his home in northern California, Ron Epstein (lecturer on philosophy and religion at San Francisco State University and the Institute for World Religions and a practicing Buddhist) explained to me how his concerns about genetic engineering are grounded in a deeper Buddhist skepticism about technology in general: "The Buddha analyzes the problems that people cause for themselves and others in terms of the afflictions that come out of a wrong understanding of self, which he categorizes as greed, anger, and foolishness. And the antidotes for these are morality, compassion, and wisdom. So for as long as there have been people, the basic motivations have always been there. The problem of technology in general is that its history is one of extending the power of these negative qualities of mind.

"Of course," Epstein noted, "you theoretically have the ability to extend not only the negative qualities but also the positive qualities of mind"[20]—but that's only likely to happen if a moral transformation occurs in people and society *before* the technology comes along; otherwise, the technology tends to continually amplify the mindset of the people who created it. It is difficult to reform the moral basis of a technology after it has been put in place. "Usually," says Epstein, "the negative effects of a technology overpower the positive effects. When it is first introduced you get all the glowing promises, when what actually happens is the flipside."

For Epstein, this general concern about the power of technology is accentuated in the case of biotech: "I think the reason there is so much concern about genetic engineering from the Buddhist point of view is that the potential for harm is on such a fundamental level with this technology and because the harm is irreversible. You have a quantum leap in the kind of short-term and long-term harm that can be done to life on the planet by this technology, even though it is fueled by exactly the same kinds of negative emotions that have been around forever."

Epstein is also bothered by the commercialization of life. "The Buddhist would say that the marketplace exacerbates negative human emotional characteristics. The economy is run by advertising systematically promoting human greed, craving, and desire. From a Buddhist point of view, this is not the path of wisdom. So we have to be careful not to use technology—particularly genetic technology—in a way that gives the marketplace power over life itself."

It Is Thy Destiny . . .

It would be a mistake to assume that all religious leaders are skeptical of biotech. Many say that the technology should be assessed on a case-by-case basis and that biotech should be seen as a gift from God to humanity rather than a Pandora's box.

Donald Munro, the Director of the American Scientific Affiliation (ASA), an evangelical Christian organization of three thousand scientists based in Ipswich, Massachusetts, is himself a geneticist. Another prominent member of ASA, Francis Collins, is head of the Human Genome Project—which a majority of Christian churches have endorsed. Collins rebukes critics of life patenting by defending it as a complex legal matter rather than a simple moral one. Still another ASA member, V. Elving Anderson, has coauthored (with Bruce R. Reichenbach) an apology for genetic engineering titled *On Behalf of God,* in which he writes that "a stewardship ethic sees technology as a gift."[21] To be sure, Anderson and Reichenbach want the developers of genetic technologies like germ-line gene therapy to exercise great care, yet they see the end as being entirely justifiable: "Should the procedures be perfected, the substantial potential technical problems circumvented, and adequate safeguards provided (no mean task, we must emphasize), germ-line intervention might begin to provide the basis for making helpful genetic changes in future generations. In this way we might be able not only to cure genetic diseases but also to enhance or develop human capacities. . . ."[22] The probable applications are already clear: "Characteristics such as tallness or fairness may or may not be desirable, and perhaps these are not worth doing research on or engineering for. But surely intelligence is another matter. Why not make our offspring smarter?. . ."[23] Making basic changes to the human genome is not without risk, but Anderson and Reichenbach believe that we should not let risk stand in our way: "Being moral agents involves risk taking. God has given to his stewards, besides the responsibilities of stewardship, the tools for making situational moral calculations, based upon moral principles."[24]

Donald Conroy of the North American Coalition on Religion and Ecology agrees with Anderson and Reichenbach that the Bible underscores humanity's role of cocreator with God. But he isn't so sure this implies a divine mandate for the development of biotech. "What does cocreativity mean? That we're working with God. But we have to remember that we don't know everything. And because our intelligence is limited we should be extremely careful as we get into things that are more and more manipulative and see what the real motives are. Is it just human domination and greed? We have to watch that we don't evoke cocreativity to justify things that we really shouldn't be doing in the first place. God's creation is good, but we humans have another side to us. We are prone to manipulate our relationships and the world around us for our own design instead of God's design. So we can easily slip from being cocreators to being destructive, from being regenerators to being degenerators."

Conroy sees the dangers of human arrogance as having deep and ancient roots: "Abraham was one of the first environmental refugees. He took his family and flocks west looking for better lands because the Sumerians had already caused ecological devastation in their area. So even though we humans are cocreators with God, we have to be very careful of our environmental destructiveness. We've made many poor choices in the past. That's why we need to adhere closely to the precautionary principle—if you're doing something that has potentially grave consequences, then you should proceed with grave caution."

Philip Heffner, a Lutheran clergyman and director of the Chicago Center for Religion and Science, takes a critical view of the motives behind the corporate purveyors of biotech, but arrives at essentially the same conclusion as Anderson regarding the implementation of the technology itself:

> I am not a Luddite. I am not opposed to this genetic science and engineering. I am not opposed to thinking about what it means for humans. I am thoroughly committed to putting as many obstacles as I can in the way of the speeding freight train of thought and action that is called COMMODITY THINKING. . . .
>
> We will continue to pursue our knowledge and technology—we have no alternative. My tradition tells me that we will do so as sinners. That means that we will fail

> to understand fully enough; we will fail to act correctly enough; we will make mistakes. The greatest fallacy-danger is to deny this and pretend that there is some perfect, paradisaical use of cloning that we will arrive at if we think clearly enough, feel sincerely enough about, and debate long enough. That will never happen. Since we are sinners and fallible, and we are also created cocreators, we ought to engineer in that fallibility-sinner factor, be as humble as hell, spend a lot of time on our knees, and recognize that if Oppenheimer thought that the atomic bomb revealed original sin, the era of genetic engineering will reveal it much more. Then as one of my tradition's mentors has said: Full speed ahead, and sin boldly.[25]

Ingrid Shafer, also of the Chicago Center for Religion and Science, uses a Jewish perspective to justify an even friendlier attitude toward biotech:

> It is important to note that awesome as recent scientific and technological "miracles" are, thus far we have done nothing except gone ever deeper into the exploration of the physical makeup of the already existent material universe. At best, we are imitating the workings of nature and learning to "play" the instrument we have been handed, as we continue to eat of the "tree of knowledge" (and now, the "tree of life"). Much depends on how we interpret the original Genesis myth. In the Jewish tradition there is no original sin, no Fall. As we eat of the tree we

> become self-conscious; we become aware of transience and our mortality; we become fully human. Ultimately this means that as formed in the Image of God, our Creator, we are meant to be cocreators (as Philip Heffner calls us) with the capacity for self-transformation. Consider the following midrash cited by Ira Progoff: "And Isaac asked the Eternal: '. . . when Thou hadst made man in Thine image, Thou didst not say in Thy Torah that man was good. Wherefore Lord?' And God answered him, 'Because man I have not yet perfected, and because through the Torah man is to perfect himself and to perfect the world.'" In that perspective, the current developments are in keeping with the original charge given to humans to be God's representatives on earth.[26]

Shafer continues: "Why is it any more plausible to imagine God erecting electric fences around certain areas of knowledge than to imagine God watching with delight and parental pride as human beings use their divinely designed brains to decipher the code of life? What's wrong with envisioning God perching on the side of a Petri dish, eager to have us correct some copyist's errors which have crept into the three billion words in the past 600 million years?"[27]

Unlike Epstein, some Buddhists have adopted an agnostic or only moderately cautious moral stance toward biotech. For its Summer 1997 issue, the Buddhist magazine *Tricycle* interviewed Buddhist scholars about cloning. Geshe Michael Roach, Abbot of Diamond Abbey in New York City, said, "It's a different mindstream. But other than that, what's the problem? I'm not sure there's a conflict with Buddhist ethics here." Judith Simmer-Brown, Chair of Religious Studies at the Naropa Institute, said, "Cloning, per se, is an interesting prospect with no real philosophical problems for Buddhists. . . ." Ravi Ravindra, Professor of Comparative Religion and Physics at Dalhousie University, opined, "Work in this field can't really be stopped. . . ." and "I am not extra worried about these new developments"; nevertheless, he admonished that "sooner or later, this will blow up in our faces. It is knowledge without conscience."

My survey of professional bioethicists, clergy, and religious scholars left me feeling less than satisfied. Not only was there lack of agreement, but few individuals seemed to have a comprehensive basis for a moral assessment of these powerful new technologies. Some were concerned only with medical issues; others were primarily interested in genetically engineered food. Few had given much thought to the implications of the technology itself or to the implications of reshaping society to conform to a discredited reductionist view of nature.

I was intrigued by Epstein's Buddhist critique of technology and the marketplace in general and could easily agree with Reigstad's comment that too few people of faith are as yet aware of the terms or stakes of the biotech debate. In contrast, Shafer's suggestion that God *wants* us to deploy biotechnologies appears naive in the extreme: even if He (or She) intends that we eat from the Trees of Knowledge and Life, does that necessarily imply that we should patent their fruit and "terminate" their seeds? Learning about living things and how they work is not the same as changing and exploiting them for short-term profit. Whether or not one believes that the events have any connection with the reputed original sin of Adam and Eve (eating from the Tree of Knowledge), it appears that *some* humans are exerting a profoundly new influence over the basic creative powers of nature and, rather than using those powers prudently and sparingly with the full consent of all who will be affected, they are recklessly rushing to apply their incomplete knowledge in a way that restricts public comment and control. The great Christian author C. S. Lewis made essentially this point when he wrote, "Man's power over nature turns out to be a power exercised by some men over other men with Nature as its instrument. . . . Each new power won *by* man is a power *over* man as well."[28]

Still, even if one believes that, in general, "Thou Shalt Not Tinker" is a safer operating guideline than "It Is Thy Destiny," the latter touches on something that begs further exploration—the "wow" that we feel when we realize that divine power is within human grasp.

CHAPTER SIX

The DNA Cult

WHETHER CALLING FOR a complete moratorium on genetic engineering or merely a go-slow assessment of specific applications, bioethicists and religious leaders have primarily addressed the "yuk" response to biotech—and understandably so. For the most part, morality is about preventing something bad from happening (hence the Ten Commandments). But what about our "wow" response? Isn't that also based in an essentially religious feeling—one that is akin to the sense of awe and mystery we feel in the presence of spiritual powers? If we are to explore and understand our intuitive responses to biotechnology, it is necessary to examine our fascination as well as our revulsion.

If the "wow" response is in fact an expression of religious awe, then we should explore the possibility that biotechnology could itself become—or perhaps already is—a type of scientific cult.

This idea takes some getting used to. Most of us associate the word *cult* with groups that claim to talk to space aliens or that sacrifice chickens or dance naked under the full moon. The meticulous laboratory regimen and the rational, precisely worded research papers of biotechnicians seem worlds removed from the popular image of cults. However, a closer look reveals some remarkable similarities.

Every cult revolves around an image or idea of the sacred; and for millions of people, genes have already acquired a quasi-religious function. In their book, *The DNA Mystique: The Gene As a Cultural Icon,* sociologist Dorothy Nelkin and science historian M. Susan Lindee explore how scientists' views of DNA are changing how people think about themselves and their world. They write:

> DNA in popular culture functions, in many respects, as a secular equivalent of the Christian soul. Independent of the body, DNA appears to be immortal. Fundamental to identity, DNA seems to explain individual differences, moral order, and human fate. Incapable of deceiving, DNA seems to be the locus of the true self. . . . [DNA has become] a sacred entity, a way to explore fundamental questions about human life, to define the essence of human existence, and to imagine immortality. Like the Christian soul, DNA is an invisible but material entity, an 'extract of the body' that has 'permanence leading to immortality.' And like the Christian soul, DNA seems relevant to concerns about morality, personhood, and social place.[1]

The idea of the soul—an immortal, immaterial essence that determines our identity and personality—has ancient roots in nearly all cultures, from Egypt to the Pacific Islands. "So, too, in contemporary American popular culture," write Nelkin and Lindee, "DNA is relatively independent of the body, gives the body life and power, and is the point at which true identity (and self) can be determined. DNA, like the soul, bears the marks of good and evil: A man may look fine to the outside world, but despite appearances, if he is evil, it will be marked in his soul—or his genes."[2]

Nelkin and Lindee's point is more than metaphorical. They are saying that DNA *has actually taken on the function of the soul* in science-based pop culture. Hundreds of novels, comic books, advertisements, and films from the past few decades all convey the idea that DNA is the essence of humanness. A typical example is a 1993 prime-time television movie called *Tainted Blood,* in which a seventeen-year-old boy from a "stable" family kills his parents, then himself. An investigative reporter (played by Raquel Welch) discovers the explanation for the shocking incident: the boy had been adopted, and his genetic mother had been confined to a mental institution. The reporter suspects the boy might have "inherited the gene for violence," discovers that he had a twin sister, and begins an urgent search for the girl with "tainted blood." But the reporter is too late and the girl commits suicide. Welch's character goes on to write a book about the incident, dedicating it to the girl, who was not to be blamed for her actions since she had been "born to kill."[3]

This iconization of DNA is directly traceable to scientists and science journalists, who often deliberately draw on religious imagery to dramatize the centrality and power of the gene, calling the genome a "Delphic oracle" or the "Book of Man" and comparing it to the Bible or the Holy Grail.

Of course, it's one thing to say that science popularizers' claims for the power of genes are overblown and quite another to say that biotechnology is a cult. In order to establish that biotech is a cult, we must first decide what constitutes a cult, then compare biotech with our description and see if it fits.

We tend to think of cults as based on simplistic but all-inclusive explanations of the world, of good and evil, and of origins that are held fervently despite contradictory evidence. Cult leaders seek to indoctrinate others with this dogma and discourage cult members from deviating from it. The ideology of genetic engineering fits this description rather well. It begins, appropriately enough, with the Central Dogma of genetic determinism, which is promoted as a simplistic and all-inclusive explanation for why people and other organisms are the way they are. Gene-based evolution explains the origin and development of life as the long-term strategy of selfish DNA molecules. DNA, as a cultural icon, offers a template for decoding the causes of good and evil in human behavior—they are the direct result of good or bad genes. Like prophets galvanizing their followers with visions of a new heaven and a new earth, genetic engineers foretell the transformation and perfection of human beings and society through the correction of nature's genetic "errors." Though biologists have produced mountains of evidence showing that the Central Dogma is simplistic and incomplete at best, its proponents continue to make ever more extravagant claims. They spread the faith through coordinated programs of indoctrination (molecular biology departments at universities) and demand the equivalent of loyalty oaths from graduate students.

Scottish bioethicist Alastair McIntosh takes this analysis a few steps further in an article titled "The Cult of Biotechnology," in which he notes that the essence of a cult is to offer (1) a deeper meaning in life (2) in ways calculated to enhance the position of cult leaders (3) while damaging alternative ways of people becoming themselves (4) with the predatory effect that life's goodness is degraded. McIntosh contends that "varying aspects of biotechnology—exercised, as it most often is, as an endeavor of capitalism in free global markets fulfill all these criteria."[4]

Biotech justifies itself as a means for the improvement of life, even to the point of offering a kind of immortality. In this case, the "cult leaders" whose positions are advanced are not only scientists but also corporate executives and shareholders motivated by desire for profit.

But how does the biotech cult damage "alternative ways of people becoming themselves"? McIntosh offers this example: If biotech can churn out food products more cheaply than ecologically benign traditional farming methods, then traditional farmers throughout the world will be at a competitive disadvantage. Ultimately, they will be forced off their land by giant agribusiness cartels that rely on gene-engineered crops and the intensive use of pesticides and herbicides. Moreover, the availability of new high-yield crops will pressure farmers to conform to world market standards and stop maintaining the less uniform but genetically diverse traditional crop varieties. "Hence biotechnology short-circuits nature and when combined with market forces, it strangles nature's way," states McIntosh.

The fourth point—degradation of life's goodness—is a consequence of the first three. If breeders promote bioengineered organisms because they are patented and profitable, the natural diversity of life will gradually be eroded. The wildness and unpredictability of nature will give way to ever greater regimen and control.

McIntosh goes on to say that the questions about whether specific genetic procedures are morally acceptable are the small questions. "Debating them is important but comprises mere displacement activity unless the big question is simultaneously addressed. The big one is so disarmingly simple and so huge that it normally remains hidden like the wood for the trees. It is: 'Whom does this serve?'"

Does biotech serve the people, especially the poor? Or does it mainly serve the financial interests of "cult" leaders? "To the Old Testament prophets," writes McIntosh, "this was the question of idolatry. Many things could be forgiven in the attempt to serve God because that service has at its essence the self-correcting qualities of love, listening, and modesty. But to serve money was, as Jesus said, to serve evil in a manner that he personified with the Aramaic word Mammon."

The Techno-Monks' Legacy

Of course, many people may object to characterizing biotechnology as a cult since, by McIntosh's criteria, other technologies could likewise be considered cultic—including nuclear reactors, internal combustion engines, and even something as basic as money. Yet we should not be too quick to dismiss this analysis just because it doesn't apply uniquely to biotech. Perhaps, rather than invalidating the critique, this objection could help us contextualize the cultic nature of biotech by placing it within the larger history of humanity's quasi-religious drive to perfect nature and itself through the use of ever-more-sophisticated tools.

Technology historian David Noble does just this in his book *The Religion of Technology,* in which he argues that "the present enchantment with things technological—the very measure of modern enlightenment—is rooted in religious myths and ancient imaginings. Although today's technologists, in their sober pursuit of utility, power, and profit, seem to set society's standard for rationality, they are driven also by distant dreams, spiritual yearnings for supernatural redemption."[5]

Noble traces these dreams and yearnings to the Christian promise of redemption and perfection, implicit in the idea that the incarnation of God (through Jesus) gave humans the opportunity to participate in God—including God's mastery over the natural world.

In Augustine's interpretation of Genesis, Adam was immortal prior to the Fall. In the Book of Revelation, Christ promises the same destiny for a redeemed humankind: "And God shall wipe away all tears from their eyes, and there shall be no more death." During the early Middle Ages, according to Noble, "technology came to be identified with both lost perfection and the possibility of renewed perfection, and the advance of the arts took on new significance, not only as evidence of grace, but as means of preparation for, and a sure sign of, imminent salvation."[6]

This development showed up first in Christian monasteries, which were the cradles of modern technology. Cultural historian Lewis Mumford, who preceded Noble by a quarter of a century in recognizing the quasi-religious nature of Western civilization's faith in technology, described medieval monastic communities this way:

> . . . from the third century after Christ in Western Europe there had been a steady withdrawal of interest from the goods and practices of 'civilization,' accompanied by a wholesale retreat from the great urban centers of power, like Rome, Antioch, and Alexandria. Little groups of mild, peaceable, humble, God-fearing men and women, from all classes, withdrew from the noisy tumult and violence of the secular world, to establish a new mode of life, dedicated to their soul's salvation. When organized as communities, these groups introduced into the daily routine a new ritual of ordered activity, a new regularity of performance, and a measure of accountable and predictable behavior hitherto unattainable.[7]

It was this effort toward "regularity of performance" and "predictable behavior" that led to the invention of clocks. At the same time, so as to have more time and energy for meditation and prayer, the monks began to devise labor-saving devices of many kinds, including the horse-powered treadmill, the watermill, and later the windmill. Soon, mechanization itself was being seen as a path to redemption.

John Scotus Erigena, court philosopher to Charlemagne's grandson Charles the Bald, wrote that the "useful arts" are "man's link with the Divine, their cultivation a means to salvation."[8] In Erigena's view, the development of technology is inherently virtuous, restoring humans to the condition they enjoyed before Adam and Eve ate of the forbidden fruit.

The Reformation greatly stimulated Europeans' utopian yearning to bring heaven down to earth. Through the conquests of Columbus and his eager followers, paradise had become an actual, known geographical place—the New World. The recovery of Adamic dominion over nature and the creation of a new Eden were widely seen as one and the same activity: a divine charge. The proliferation of new transportation and manufacturing techniques went hand-in-hand with the plunder of exotic lands.

While clocks and watermills had originated in monasteries, their ability to increase profits by reducing human labor was soon recognized by the merchant class, who enthusiastically adopted them and put them to a wide range of uses. But the religious and idealistic vision that had spurred the European technological project remained in place, granting the proceedings a moral imperative.

King James's Lord Chancellor, Francis Bacon, defined the Western project of modern technology perhaps more forcibly than anyone else before or since. "Truth and utility are the very same thing," he wrote.[9] In his utopian novel *New Atlantis,* Bacon imagined a guild of scientists called Solomon's House that had reclaimed their rightful dominion over the earth and thereby brought about a restoration of paradise. Possessed of "the knowledge of causes and secret motions of things," the members of Solomon's House were also capable of "the curing of diseases counted incurable . . . the prolongation of life . . . the transformation of bodies into other bodies [and] the making of new species."[10]

Thus, even in its earliest stages, the Western technological project included a vision of what would later become genetic engineering.

Perfecting Nature, Perfecting Ourselves

To relieve any doubts that biotechnology feeds on cultic visions and motives, one need only consider the story of eugenics.

The term *eugenics* was coined by the British biologist Sir Francis Galton (a cousin of Charles Darwin) in 1883. He defined it as "the study of the agencies under social control which may improve or impair the racial qualities of future generations physically or mentally"—the way plant and animal breeders improve their stock. In essence, eugenics means the application of scientific breeding techniques to human beings. It represents, in some respects, the ultimate achievement of the religion of technology in its quest for transcendence and transformation.

In America, the eugenics movement thrived between 1900 and 1935 and was promoted by famous figures including plant breeder Luther Burbank, inventor Alexander Graham Bell, Stanford University president David Starr Jordan, and industrialist John Kellogg—and, as we have already seen, the Rockefellers and their Foundation's board members.

Eugenics was not a single coherent doctrine and was not

promoted by a single group; rather, like the "gene mystique" of today, it was taken up by many seemingly unrelated organizations and individuals including Baptist ministers, labor unionists, pig breeders, and elementary-school teachers. It offered, in Jeremy Rifkin's words, "a scientific explanation for social and economic problems and a scientific approach to their solution."[11]

Eugenicists were emotionally committed to a vision of a human future free of disease and criminality. Just as no cattle breeder would seek to perpetuate the traits of his worst animals, so no society should allow its weakest citizens to procreate. Popular literature spoke of the "born criminal," and in 1917, science writer T. W. Shannon proclaimed that the "hereditary nature of the criminal propensity is unquestionable."[12]

Throughout the early decades of the century, eugenics functioned as a kind of civil religion. As priests of that religion, geneticists offered conceptions of both heaven (a future of universal health and intelligence) and hell (the alternative future of universal disease, mental illness, and criminality that would ensue if eugenic measures were not adopted). Sociologists devoted research projects to the study of dysgenic families through many generations, in the belief that genes were the key to understanding every form of human depravity. Meanwhile, eugenicists clung to the utopian ideal of a future race of Shakespeares and Newtons that could result if appropriate action were initiated. National conferences on "Race Betterment" were convened, at which prizes were offered for "better babies" and "perfect school children," based on physical measurements and mental tests.

Soon the fervently held beliefs of eugenicists were translated into action: states began passing compulsory sterilization laws, leading to the involuntary sterilization of tens of thousands of American citizens. Indiana passed the first such statute in 1907; it called for the mandatory sterilization of "confirmed criminals, idiots, imbeciles," and others housed in state institutions, upon approval by a board of experts. Ultimately, twenty-nine other states would pass similar laws. In Missouri, sterilization could be imposed not only on murderers and rapists, but on those convicted of "chicken stealing" and "theft of automobiles" as well. The constitutionality of the sterilization laws was upheld when, in the 1927 Supreme Court case *Buck v. Bell,* Justice Oliver Wendell Holmes approved the forced sterilization of a teenage mother and her young child on the grounds that "three generations of imbeciles are enough."[13]

By the mid-1930s, however, concerns were being raised that the practice of eugenics—however scientifically justifiable—would inevitably be subject to corrupt administration within an imperfect social system. A 1934 film entitled *Tomorrow's Children* tells the story of a state welfare worker who seeks to sterilize a hardworking, attractive, and kind young woman from a dysgenic family. Meanwhile, a corrupt and violence-prone man from a wealthy family is spared the knife because of his father's political influence. The young woman likewise escapes sterilization only when it is revealed that she had been adopted as an infant.

Of course, the eugenics movement didn't exist only in America. It reached full flower in Nazi Germany, where a state-sponsored "racial hygiene" movement carried out forced sterilizations on a massive scale. In their promotion of the ideal "master race," the Nazis finally turned from sterilization to extermination.

It isn't enough to say that eugenics influenced Nazi policy; it would be more accurate to say that German policy leaders regarded National Socialism as applied biology. Gerhard Wagner, leader of the German medical profession during the Nazi era, wrote that "knowledge of racial hygiene and genetics has become a purely scientific path, the knowledge of an extraordinary number of German doctors. It has influenced to a substantial degree the basic world view of the State and indeed may even be said to embody the very formulations of the present state. . . ."[14]

After the war, revulsion at Nazi "racial hygiene" experiments resulted in the suppression of the eugenics movement in America and most other countries. Through the 1950s and 1960s, the eugenic philosophy survived in only two organized scientific fields—human genetics and infertility research. But as these two fields grew in influence during the latter decades of the century, eugenics gradually made a stealthy comeback. Today the sorts of items one commonly reads in the newspaper about the power of genes to influence everything from homosexuality to IQ to obesity closely resemble statements made by leaders of the eugenics movement of the 1920s.

However, the new eugenics appears to have little to do with oppressive state coercion; rather, it seems more a matter of individual choice motivated mostly by parental love. The manner in which this kinder, gentler eugenics is likely to be applied was previewed in a 1984 short story by Rena Yount, entitled "Pursuit of Excellence." Set in a future society, the story portrays wealthy parents who buy genetic engineering services in order to endow their children with spectacular physical beauty and intelligence. These rich designer children grow up to run the society, holding all the top positions in government and industry.[15]

Yount's "Pursuit of Excellence" is still in the category of science fiction, though perhaps not for long. The story hinges on the development and use of germ-line gene therapy. By all accounts, this technique is still years from implementation in humans, and many bioethicists (not including Arthur Caplan) argue against its ever being tried. But equivalent procedures are already being applied to farm animals and, given the pace of current developments, it is reasonable to assume they will soon be tried on humans. Certainly the motivation is not lacking—what caring, wealthy couple would withhold from their children a procedure that could grant them immunity from genetic diseases or endow them with better looks or more intelligence?

Caplan insists that comparison of germ-line therapy with Nazi eugenics is pointless. Then, eugenics served racist and political purposes; gene therapy is for individual betterment. "There is no slope that leads inexorably from therapeutic germ-line interventions intended to benefit future persons to the creation of eugenically-driven, genocidal social policies," he writes. "Nazi eugenic policies were not aimed at benefiting individuals. The state or the Volk, not the individual, was the object of Nazi eugenic policy. Public health, not individual therapy, was the driving force behind the Nazi medicalization of eugenics."[16]

Colin Gracey disagrees. "There are many who say that today's eugenics is being approached for different reasons, from a different knowledge base. But to me that misses the point. My concern is that, at any moment in history when people in power embark on a biological approach to social engineering, that's inherently a value-laden enterprise. I see no way of preventing the decision makers from projecting their values into the process. I don't see how you could avoid a situation that would be ethically and spiritually equivalent to what the Nazis were doing, even if it didn't entail mass sterilizations and extermination. In my view, whatever faith community you come from, you have long-term interests in truth, freedom, and justice. And those get preempted with eugenic thinking."

Margaret Mellon of the Union of Concerned Scientists sees even the effort to find the genetic roots of diseases as tending to lead society toward eugenic thinking. "You see it in the newspaper all the time," she told me. "Scientists find that genes are associated with a particular disease, and all sorts of conclusions are drawn before anybody has an idea whether the data are any good or not. The people who suffer from the disease really want to have a cure, but then there are also economic interests in being able to come up with new treatments for disease, and that leads people to use a kind of shorthand which at times is not utterly incorrect but which I think heads off in the wrong direction. People say things like, 'It's because you have *x* gene that you suffer from *y* syndrome.' That may be partially but not completely true, and to the extent that people think it's truer than it is, it has distorting and destructive implications. If you go back and look at eugenics, a lot of the theories and their horrific applications came out of an overemphasis on the role that our genetic constitutions play in who we are. On a macro-political scale, this kind of thinking leads people not to want to improve the conditions that folks live in. For example, rather than reduce the load of pesticides and other harmful chemicals on its employees, a company might instead decide not to hire people who are genetically more susceptible to pollutants."

Today, the gene cult has already profoundly influenced the way we think about ourselves and the world around us. Increasingly, responsibility for social ills is being attributed to DNA—thus absolving the family environment and society itself. More and more often, we look to genes for the answers to complex problems. Criminals and geniuses, saints and sinners, the weak and the strong, and the diseased and the healthy are the way they are because *they were born that way*. Why make the effort to improve the social context in which people are born and raised when it would be simpler and more effective merely to alter a few genes? We have yet to ask a logically related question: Rather than expecting our leaders to attain a highly cultivated moral sensibility, why not just clone a Jesus or a Buddha?

Beyond Hubris

As we've discussed, there are essentially three moral or religious responses to biotech, which we might characterize as: Yuk (Stop! It's wrong!); Yuk/Wow (In some ways it's good, in some ways it's bad; it all depends); and Wow (Go for it!).

Most bioethicists are inclined to take the middle path—to say that biotech has the potential for both good and bad and that by discussing our options in the light of ethical principles we can learn to take the former and leave the latter. However, while this response may motivate us to assess specific applications of the technology, it may not help us see or understand the subtle effects on society of the technology itself. Those effects—on our thinking, our values, and our relationships with one another and with the natural world—are often difficult to sort out except in hindsight. Moreover, this approach fails to appreciate the quasi-religious nature of the enterprise and the fact that an entire culture can be swayed by cultic visions of social transformation through genetic purification.

It is, of course, important to engage in discussions about specific applications of biotech and arrive at some collective agreements about which ones are simply too cruel, destructive, or dangerous and which ones entail acceptable levels of risk to the environment and ourselves. However, since the technology by its very nature is bound to have cultural effects regardless of our microethical decisions about it, we must also reflect critically on the technology as a whole. So although it is possible to oppose some uses of biotech and not others, at the metalevel of technology assessment we are still left really with two fundamental responses—either skepticism or embrace. Yuk or wow.

These responses reflect two faces of spirituality, both with ancient origins. One is the spirituality of humility and respect, the other the spirituality of power and transcendence.

During the last few centuries, in Europe and America, the spirituality of transcendence has clearly won the day. If the goal of technology was originally to improve the condition of humans, the goal of human life has, in Mumford's words, "become ever more narrowly confined to the improvement of technology."[17] The equation of mechanical progress with moral progress, implanted in the Western mind in the late Middle Ages, has become an unconsciously held truism. And it has led to absolute wonders of technical virtuosity: we have blasted off the planet into outer space, we have split the atom, and we have transformed the world in a million ways to suit our convenience and comfort.

Hence the argument so often heard for biotech: "We've already changed plants and animals through breeding; we've already redesigned the natural world. Why stop now?"

Not only humans but many other creatures as well seek to adapt the environment to themselves, as well as adapting themselves to their environment. Beavers build dams; birds build nests. We humans have hewed to the former strategy about as single-mindedly as it is possible for any creature to do. We have already remade much of the world to suit our short-term desires. There are plenty of signs that we have already gone too far in that direction and that we would do well to put more effort into adapting ourselves to suit nature. Perhaps, therefore, it is high time that we revisit the spirituality of respect and humility.

CHAPTER SEVEN

A Different Path

THROUGHOUT MY RESEARCH on the moral impact of biotechnology, I have kept a growing file of the basic arguments, pro and con. I've already discussed most of these arguments at some point in this book, but to assess further the current state of the public debate, I will summarize them here.

Usually, ethical overviews of biotech sort the philosophical issues according to various technical applications—gene-altered food, somatic gene therapy, germ-line gene therapy, genetic screening, and so on—and list pro and con arguments for each. While that approach is valid, it tends to gloss over the manner in which all biotech applications flow from a single, genetic-determinist worldview. Therefore, I've chosen a different manner of presentation, one that catalogs arguments according to their ideological basis.

Against Biotechnology: Moral Arguments

A. The Spiritual Argument

Genetic engineering arises from, and reinforces, the basic assumption that an organism is an *object* to be manipulated rather than the *subject* of its own experience and determiner of its own destiny; it is the ultimate expression of a worldview based on control and dominance.

This utilitarian view of organisms is not entirely new; we humans have controlled nature in varying degrees throughout our history. And (again to varying degrees in different cultures) we have looked upon other organisms, and other people, as objects to be manipulated. Human slavery is perhaps the most abhorrent practice to come from this perception of the world, while the meat and dairy industries offer examples to which many people still do not object. But with genetic engineering, the tendency to objectify organisms and persons—avoiding empathetic responses—culminates, since we are proposing not only using the organisms themselves for our purposes but also permanently altering the genetic code (again, for our purposes) of all their future progeny. Moreover, by contemplating the modification of our own genes, and by growing genetically matched "spare parts" for organ transplants, we are treating even our own bodies as technological artifacts.

In its extreme, the utilitarian view of nature becomes a kind of Promethean cult whose directives are to transcend material limits and transform ourselves and to control and manipulate nature in the most absolute terms possible. In this cultic vision, *we* are the creators of life and cosmos, and our ultimate goal is to recreate ourselves.

In the modern world, Prometheanism functions as a secular religion administered through the triune powers of science, technology, and free trade. Science promises to empower us to *know* everything, technology promises to empower us to *do* everything, and free trade promises to empower us to *buy* anything. Any suggestion that these three powers be leashed to some higher public good is regarded by cult defenders as tantamount to blasphemy.

Biotechnology, with its implicit commodification of the very essences of life, forces spiritually oriented people to ask these vital questions: Is spirituality only about becoming gods? Or is it equally—even primarily—about confirming our connection with the rest of life and acknowledging our humility before powers higher than ourselves?

B. The Religious Argument

The adherents of several religions have more specific objections to some aspects and applications of biotechnology. Members of faiths that specify dietary guidelines object to crossing species barriers through the genetic engineering of foods. Other people of faith have more general or subtle objections: many Christians, Jews, and Muslims believe that genetic engineering violates God's plan for living things, while some Buddhists believe that exercising our technological ability to engineer genes only serves negative human motives such as greed and the desire for power.

C. The Scientific Argument

Biotechnology assumes a reductionist, mechanistic, and determinist attitude toward genes and life. As we saw in chapter 1, the assumption at the core of genetic engineering and cloning is that organisms are discrete objects and that their DNA is like a computer program determining the organism's traits in a straightforward and predictable way. Genes thus become interchangeable parts, like the parts in a machine made on an assembly line.

The evidence, however, indicates that the situation is far more complex and that the effects of any specific DNA sequence are qualified by the rest of the genome, by the organism as a whole, and by the relationship between the organism and its environment.

Research based on reductionist assumptions has produced some remarkable experimental successes—organisms that appear to have precisely the characteristics we want engineered into them. However, it has also generated many mysterious failures. This is especially true in research with multicelled organisms, in which the desired characteristic sometimes doesn't show up, is accompanied by somatic or behavioral changes, or has unexpected effects on ecologically related organisms.

By failing to take context into account, scientists are radically oversimplifying their view of genetics. Their tendency to do so may be tied to the economic and political influences of immensely profitable and powerful chemical and pharmaceutical conglomerates that pay the salaries of so many research scientists.

Sometimes even small errors in our assumptions can lead to serious problems. For example, the belief of the *Titanic's* designers that the ship was "unsinkable" was a factor in their decision to carry fewer lifeboats than needed to accommodate all the passengers. In the case of biotech, the error in thinking may be just as deep seated and profound and the ensuing problems just as catastrophic. We are already experimenting with the human food supply and the ecology of the natural world, and we are preparing to experiment with the genetic makeup of future generations of human beings.

Therefore, many prestigious scientists—including Lynn Margulis, David Suzuki (geneticist, television journalist, and author of *Genethics*), Ruth Hubbard (Professor Emerita of Biology, Harvard University), and Erwin Chargaff (Professor Emeritus of Biochemistry, Columbia University)—have called for a far more cautious approach to the application of genetic science than is presently being pursued. Chargaff, often referred to as the father of molecular biology, once described genetic engineering as "a molecular Auschwitz" and warned that it poses a greater threat to the world than nuclear technology. In his autobiography, *Heraclitean Fire,* he wrote: "I have the feeling that science has transgressed a barrier that should have remained inviolate."[1] Dr. George Wald of Harvard, a Nobel Laureate in the field of medicine, has written that "Recombinant DNA technology faces our society with problems unprecedented not only in the history of science, but of life on earth. . . . Up to now, living organisms have evolved very slowly, and new forms have had plenty of time to settle in. Now whole proteins will be transposed overnight into wholly new associations, with consequences no one can foretell, either for the host organism, or [its] neighbors. . . . It is all too big and is happening too fast. So this, the central problem, remains almost unconsidered. It presents probably the biggest ethical problem that science has ever had to face."[2]

D. The Social Argument

Some of the ethical questions raised by biotech are old and familiar—including those regarding the concentration of wealth and power within society, and society's distribution of the unequal economic benefits and costs of technological change.

Typically, when a powerful new technology is introduced into a society that is already inequitable, implementation of the technology will exacerbate the gap between rich and poor. We saw this happen in America during the last two centuries with the introduction of railroads and various petroleum-based industries. On one hand, fortunes were made; on the other, whole communities were systematically exploited and impoverished. Today we see the same phenomenon occurring throughout the world as corporations seek to lay claim to the genetic heritage of the planet. Huge profits go to agricultural and pharmaceutical companies, while the cultures and communities from which genetic "raw materials" are extracted receive little or nothing in return and are, in fact, placed at an increased economic disadvantage.

It usually takes at least a generation to see the full range of effects from a new technology. This was true of automobiles, nuclear energy, and synthetic chemicals; it will likely be true of biotech as well. This is to say, not that all the fallout from genetic engineering will be bad, but simply that *we don't know* what it will be. All we can say for sure is that casualties are probable. Who will pay for whatever damage ensues—corporations, laboratories, or society as a whole?

Margaret Mellon emphasized in our conversation that scientific and technological choices are also social choices. "Once you know that there are lots of organisms on earth and you have the power to manipulate them to achieve some end, you have to ask whether that's the right use of your knowledge and your technology. It's certainly not the only way to arrive at a variety of desirable end points, and yet it's favored because it dovetails so completely with our economic system. If your problem is pests, and you solve that problem by creating a pesticide or a plant that can repel pests, then you've got a product you can sell to folks, you've got an income stream, you've got a company and a job and a reason to do science. If, on the other hand, you solve the pest problem by rotating existing crops, then you still come up with a solution—but since you haven't created a new product, you don't create a new income stream, you don't create new jobs, and all of these other things.

"So the availability of biotechnology, particularly in our economic system, biases our approach to many problems," says Mellon, "and that has distorting effects on our scientific establishment and on the kinds of economic activities we pursue. For example, it tends to favor reductionist science, which breaks organisms down into their component parts to study them, rather than studying how organisms interact with one another in ecosystems. There's a reinforcing quality in how reductionist science leads to certain kinds of solutions, which lead to products, which influence the reasons to do science; it all leads around in a circle. The whole process leaves us seriously out of balance in the kinds of science we do and in the directions we take in solving major social problems like providing health care or producing food."

One highly probable social effect of genetic engineering as applied to humans will be the differential application of its benefits, so that wealthy children will be more likely than poor children to be screened for genetic diseases and provided with cures. The eventual implementation of germ-line gene therapy will mean the emergence of a class structure that is both economic *and genetic* in nature—a full-blown eugenic society.

Social decisions affect the implementation of new technologies. Should living things, body parts, and genes be patented for profit? Should society invest in developing biological weapons of mass destruction? These are not technical decisions to be left up to scientists, nor financial ones to be left up to entrepreneurs and corporate CEOs. They are moral questions that must be addressed by all members of society, because in the end all of society's members will have to deal with the consequences.

E. The Ecological Argument

When genetically altered organisms are released into the environment, the potential effects extend beyond human civilization. Humankind has been modifying its environment for thousands of years, and many of the effects—especially more recent ones resulting from burning fossil fuels and using petrochemicals—have been ecologically disastrous.

The potential environmental effects of biotechnology are largely unknown. We do know that the phenomenon of "escaping genes," long expected by biotech critics, has already become a reality. A 1998 government study in Germany on genetically altered rape plants (a close relative of the turnip, grown commercially for "canola" oil) reported that the plants passed on their herbicide-resistant gene to ordinary rape growing as far as two hundred meters away. Monika Griefahn, the Minister of Environment in Lower Saxony, where the study was carried out, said that this result confirmed her worst fears: "Once the manipulated genes are released into the surroundings, there is no way to contain them."[3]

Other unintended ecological consequences are likely. Not only are organisms highly complex, but they depend upon one another in natural ecosystems in ways that we do not fully understand. Scientists may intend the addition or deletion of a few genes in an organism to have a single, specific effect, yet an apparently minute change may have incalculable consequences not just for the engineered organism but for other species as well.

The effects on pollinator species alone are worth noting. French researchers studying rape plants with transplanted fungus-resistance genes discovered that the plants posed a threat to nearby colonies of bees. After two weeks' exposure to the plants, the bees' memory for smell—the basis for their ability to locate food sources—was dramatically reduced. Elsewhere, preliminary studies have shown a 30 percent drop in bee population in areas where genetically engineered cotton was being tested.[4] Industrial agriculture already poses a serious threat to the populations of many essential pollinator species (including flies, bees, butterflies, bats, and birds). Further reductions in the number of pollinators could in turn disrupt agriculture, resulting in a threat to the human food supply—in addition to having a catastrophic effect on noncultivated flowering plants.

David Letourneau expressed the ecological argument clearly in our interview: "I believe that genetic engineering is ultimately a worse threat than nuclear or chemical pollution. Radioactive elements have a half-life; chemicals break down. But genes will go on and on. Once they're in the environment, there's no containing them or calling them back."

F. The Argument Against Unnecessary Cruelty

This final argument applies only to the applications of biotechnology having to do with animals. A great many biotech experiments on animals are extremely cruel. For example, in a recent study in which USDA researchers attempted to create a big, fast-growing pig with superior meat, they injected eight thousand pigs with a human gene governing growth. Of these, only one produced the desired hormone. However, this one "successful" pig (#6707) was excessively hairy, lethargic, arthritic, impotent, and cross-eyed and could hardly stand up. The other 7,999 were "failures" and were disposed of accordingly.

The number of animals used in such experiments is not trivial (actual figures are hard to obtain, but estimates run into the hundreds of thousands). Even when experiments "succeed"—as with DuPont's patented "oncomouse," engineered to develop cancer as an aid in human cancer research—ethicists still object to treating animals purely as means to human ends. Dr. Michael W. Fox, Vice President of the Humane Society of the United States, writes:

> It is morally wrong to violate the right and entitlement of animals to humane treatment. This ethic is written into law, as witness the federal Animal Welfare Act and state anticruelty statutes. Since the genetic engineering of animals may cause them to suffer from physical and physiological changes . . . it is surely unethical and a violation of humane ethics and legal statutes to subject animals to such manipulation. . . .[5]

For Biotechnology: Moral Arguments

As far as I can tell, there are two main moral arguments *in favor of* the continued development of biotechnology and against any effort to stop its development:

A. The Argument Against Repression

Many people regard the freedom of scientists to engage in an unimpeded search for knowledge as quasi-sacred. It is a freedom that was gained at considerable cost (recall that Giordano Bruno was burned at the stake by the Inquisition in 1600 for insisting that the Earth orbits the sun). If genetic research were forcibly halted for essentially religious reasons, then a precedent would be established. Eventually *all* scientists might have to answer to councils of citizens authorized to decide what basic knowledge the rest of us should permitted to obtain and transmit. Instead of a clear, objective understanding of the universe, we would come to have one that was slanted according to the desires of the group appointed as the moral watchdog of society. The "creation science" taught in fundamentalist Christian schools is only a preview of what all of science might become should it evolve in such a biased and constricted social context.

B. The Argument Against Depriving Society of Potential Benefits

It is not unreasonable to think that genetic research could eventually find cures for cancer, AIDS, Alzheimer's, and many other horrific diseases. Proponents argue that it could also lead to increased agricultural outputs, helping feed the burgeoning human population of our finite planet. There may be other benefits as barely yet imagined.

If biotech research can cure diseases or feed starving people, wouldn't it be morally wrong *not* to pursue it?

❖ ❖ ❖

By this point in the book, it is certainly apparent that my personal assessment of biotechnology and most of its applications tends to be skeptical and negative. That's because, overall, I find the arguments for strictly controlling biotech to be stronger and more convincing than the arguments against doing so. But, having come to this conclusion, what am I to do with the two arguments just cited—that such controls would constitute an infringement on scientific freedom or that society would be denied important benefits?

The first can be answered in two ways. First, few antibiotech activists are actually calling for the curtailment of basic genetic research, except where the research process is itself inherently unethical or inhumane (as is the case in many experiments involving animals). What many people do object to is the headlong rush to *apply* knowledge gained from genetic research before fully understanding its implications.

The second, more philosophical argument assumes that the search for scientific knowledge is inherently value free, but this is not necessarily true. There is an infinite amount of knowledge about the universe that is available to be uncovered. Where shall we look? What knowledge shall we seek to acquire? For example, shall we seek to gain knowledge that will enable us to live in harmony with nature, or shall we seek knowledge that will enable us to dominate and manipulate nature?

Many philosophers and historians of science believe that knowledge is seldom purely "objective"; it always proceeds from, embodies, and promotes a particular set of values. As anthropologist C. R. Hallpike puts it, "The kinds of representation of nature . . . that we construct [flow from the way] we interact with the physical environment and our fellows."[6] A few biotech critics believe that, in addition to placing a moratorium on many applications of the technology, we should also examine the direction of our research. If genetic research is being guided by reductionist, utilitarian, and mechanistic assumptions in a social context that is slanted and constricted by corporatism, then our assumptions must be reevaluated and science must be freed from the imperatives of corporate profit. Until some effort has been made in those directions, it is pure hypocrisy to assert that current genetic research is "objective" and that citizen oversight would necessarily constitute an infringement of scientific freedom.

The second argument against controlling biotech is just as slippery. How do we weigh the harms and benefits of genetic engineering? It's difficult to do so when the benefits are constantly being touted by well-paid industry spokespeople, while the harms may not show up for a generation or so and the people warning of them tend to be poorly funded activists. Suppose we cure one disease but in the process create another that is even worse. Suppose we engineer a short-term increase in the human food supply but undermine the ecological basis not only of our food system but of the entire biosphere. Biotech boosters say that such hypothetical concerns reflect undue pessimism, but the history of the introduction of new technologies suggests that pessimistic forecasts of technological side effects are often at least as accurate as proponents' rosy promises.

A thorough discussion of benefits and harms likely to flow from biotechnology would fill a book much longer than this one. It's worth examining in more detail at least one proposed benefit—the promise of increased food production through genetically altered crops. In promotional literature, Monsanto claims that "biotechnology innovations will triple crop yields without requiring any additional farmland, saving valuable rainforests and animal habitats," and all this with "less chemical use in farming."[7] Are these promises credible?

Monsanto's claims may be based on several basic misconceptions. The first is that hunger is caused primarily by a shortage of food with which to feed a growing population. Yes, world hunger is a real problem. But it is not caused by lack of food.

Frances Moore Lappé, Joseph Collins, and Peter Rosset have shown in their book *World Hunger: Twelve Myths* that the world already produces enough grain to provide every human being on the planet with 3,500 calories a day. This estimate does not take into account many other commonly eaten foods, such as vegetables, beans, nuts, root crops, fruits, grass-fed meats, and fish. Altogether, there is enough to provide at least 4.3 pounds of food per person a day. Hunger persists despite the fact that increases in food production during the past thirty-five years have outstripped the world's population growth by about 16 percent. According to Dr. Peter Rosset, Executive Director of Food First, "The true source of world hunger is not scarcity but policy; not inevitability but politics. The real culprits are economies that fail to offer everyone opportunities and societies that place economic efficiency over compassion."[8]

Andrew Kimbrell, President of the International Center for Technology Assessment, argues that one of the greatest causes of world hunger is food dependence—the fact that increasing numbers of people do not grow their own food but must purchase it with scarce cash. Food dependence comes from industrial society's enclosure of land, which forces peasants off the land so that it can be used for export crops. "The result of enclosure," writes Kimbrell, "has been, and continues to be, that untold millions of peasants lose their land, community, traditions and most directly their food independence."[9] Flocking to cities, former peasants become the new urban poor. But are the megafarms that displaced them really so efficient? According to Kimbrell, "Conventional efficiency analysis completely ignores the social and environmental cost of large-scale industrial farming. The costs of water and air pollution, topsoil loss, biodiversity loss are not considered"—nor are the human health costs of eating pesticide-laced food or the social costs stemming from the millions of dislocated peasants. None of these costs are factored into the "real" price of food produced on industrial farms. In 1989 the United States National Research Council was commissioned to assess the efficiency of large industrial farms as compared to smaller scale alternatives; their conclusion was that "well-managed alternative farming systems nearly always use less synthetic chemical pesticides, fertilizers, and antibiotics per unit of production than conventional farms. Reduced use of these inputs lowers production costs and lessens agriculture's potential for adverse environmental and health effects without decreasing—and in some cases increasing—per acre crop yields. . . ."[10]

Biotechnology builds on the failed assumptions at the core of industrial agriculture. As agricultural biotech takes hold, farmers are becoming ever more dependent on giant corporations like Monsanto—not only for pesticides and fertilizers (of which they will eventually need more, not less), but also for patented, gene-engineered seed. The enclosure of the biological commons will soon be complete. And should the centralized, industrial food system ever fail for any reason (i.e., natural disasters, corporate mismanagement or financial woes, or the unexpected ecological side effects from altered genes), the present simmering crisis of world hunger could explode into world famine.

Would foregoing this "benefit" to society really constitute a moral problem?

The Biotech Wars

Often when I've tried to discuss these issues with students, friends, and colleagues, they've expressed skepticism that controlling biotech is actually possible. One of the comments I've heard most frequently is, "The genie's already out of the bottle. You couldn't stop biotechnology now even if you wanted to."

This attitude, it seems to me, is more a result of Americans' general frustration with politics than of the realities of the new genetic technologies. The fact is, even though the activists who are working to control the implementation of biotech are mostly ill funded and overworked, *they are already making a clear difference.*

Many grassroots antibiotech groups engage in direct action, making up in courage and humor for what they lack in financial backing. In 1996 Greenpeace activists in the American Midwest sprayed a red X on an entire crop of Monsanto's genetically engineered soy beans. "These are the X fields," declared Greenpeace, alluding to the frightening conspiracies of television's *X-Files*. "We are bearing witness to a huge genetic Xperiment being carried out on the global environment." Over the next few months, Xs appeared at the headquarters of Nestle, Danone, and Unilever, were applied to margarines and other soy products in supermarkets throughout Europe, and were sprayed on the sides of seed ships containing genetically engineered grain.

In March 1997, the Women's Environmental Network painted a bar code on the stomach of a pregnant activist and picketed lawyers who were engaged in patenting human umbilical cells for the biotech firm Biocyte. A month later, Earth First! members, animal rights activists, and radical organic farmers held a Big Gene Gathering in Herefordshire, England to plot the overthrow of "the biotech transnational corps(e)." Only days after that conference, fifty costumed "Super Heroes Against Genetix" (SHAG) took over the boardroom of Monsanto's U.K. headquarters, engaged management in discussions about genetic pollution, and hung a banner from the roof declaring "Gene Wars—the consumer strikes back!"

In Carlow, Ireland, the Gaelic Earth Liberation Front (GELF) spent a November night in 1997 digging up a Monsanto genetically altered sugar beet crop and afterward boasted that they had made Ireland "genetech-free" again. In Germany, activists camped out in fields scheduled to be planted with genetically altered crops and held genetech-free farmers' markets there. And in January 1998, one hundred farmers from the second largest of France's farmers' unions, Confédération Paysanne (CP), entered a Novartis conditioning and storage plant, found five tons of transgenic maize, and destroyed it. Three of the farmers were arrested; outside their court hearing, a thousand people gathered in support.

Not all antibiotech actions consist of street theater. Some represent serious legal challenges to corporations' "rights" to patent life forms or to carry on potentially dangerous research and trade. In the U.S., Jeremy Rifkin's Foundation on Economic Trends (FET) has on several occasions organized coalitions of citizen groups to oppose various aspects of biotech research and has filed legal briefs to halt certain biotech procedures. In 1985 the FET formally requested that the NIH temporarily suspend funding certain transgenic animal experiments pending a review of the ethical implications. In 1994 FET assembled a coalition of hundreds of women's organizations from more than forty nations to oppose Myriad Genetics' attempt to patent a gene that causes breast cancer. And in 1995, FET organized a coalition of more than two hundred religious leaders—including the titular heads of nearly every major Protestant denomination, more than one hundred Catholic bishops, and Jewish, Buddhist, and Hindu leaders—to announce their opposition to patenting animal and human genes, organs, and tissues.

In March 1998, European scientists and doctors at a special hearing in the European Parliament called for a moratorium on genetically modified organisms (GMOs). They explained that the long-term impact of GMOs on human health and the environment has not adequately been assessed and cited the "precautionary principle" adopted by more than one hundred Heads of State and Government at the Rio Earth Summit in June 1992. The doctors and scientists claimed that public debate has been dominated by pressure from the biotech industry, while public interest has been ignored.

And on May 28, 1998, a coalition of scientists, religious leaders, health professionals, and chefs filed suit against the FDA, claiming that the failure to label thirty-three different genetically engineered foods violates the agency's mandate to protect public health and provide consumers with relevant information about the food they eat. Among the plaintiffs were Donald Conroy, Colin Gracey, Ron Epstein, and Philip Royal.

Efforts such as these are producing results.

• In 1996 nineteen European nations signed a treaty that called cloning people a violation of human dignity and a misuse of science. (Britain and Germany balked at signing the measure because London considered it too strict and Bonn too mild.) The same day, French President Jacques Chirac called for an international ban on human cloning, and two days later President Clinton urged Congress to do the same.

• In 1998 Iceland Foods, a major food chain in Great Britain, announced that it would refuse to use any genetically modified ingredients in its line of food products. Malcolm Walker, Iceland's founder, chairman, and CEO, has said that genetically altered foods represent "probably the most significant and potentially dangerous development in food production this century." A pamphlet issued by the food chain, "Genetic Modification and How It Affects You," outlines many of the unknown risks of the new technology and urges "all our customers, other retailers, manufacturers and farmers to join our campaign for crop segregation and a more cautious approach to the introduction of this new technology."

• The government of Thailand moved in November 1997 to call off field tests on Bt cotton engineered to kill insect pests. Thai officials feared that foreign toxin genes could spread to related plants and kill insects in wider areas of the country. Also, makers of Thai herbal medicines were concerned that the bioengineered cotton would cross-breed with other species of the cotton family, which are used in the production of traditional health remedies, and thus interfere with their medicinal properties.

• In 1997 Dutch authorities disposed of twelve thousand tons of sugar after discovering that it was contaminated with sugar from genetically altered sugar beets. The following year, the British sugar beet industry refused to accept genetically engineered sugar beets. Both moves came in response to growing public opposition to genetically engineered foods. A spokesperson for British Sugar, Geoff Lancaster, was quoted as saying that "public suspicion may sink this technology completely."

• Early in 1998 the USDA proposed a set of national standards for organic foods. The new standards would have allowed genetically altered varieties—if grown without pesticides or herbicides—to be labeled "organic." USDA officials invited public comment and were promptly inundated with letters, faxes, and E-mails. During the next ninety days, more than two hundred thousand individuals and organizations submitted comments, the vast majority of which disapproved of the new rules. In April 1998 the USDA retracted the rules and proposed a new set excluding genetically altered foods.

• In Austria, environmentalists, farmers, food producers, and grocery store owners have banded together to create a nationwide safety seal guaranteeing that foods do not contain genetically modified ingredients. Also in that country, grassroots efforts have forced Pioneer Seeds to abandon its efforts to grow pesticide-resistant corn. As a result of public pressure, there were no genetically engineered crops grown in Austria in 1998.

• In May 1998 the European Union adopted a law mandating the labeling of food containing genetically altered ingredients. The statute came in response to complaints from consumers still reeling from fears generated by Europe's mad cow crisis. However, environmental groups warn that a majority of gene-altered products may not be covered by the rule, since in many cases the genetic manipulation cannot be traced. The U.S., a major supplier of corn and soy to Europe, does not segregate conventional and gene-altered crops in bulk deliveries.

Naturally, the gigantic, well-financed biotech industry is not taking developments like these lying down. Agricultural, medical, and pharmaceutical biotech corporations pay high-priced advertising and public relations firms, lobbyists, and lawyers to advance their interests at every opportunity. And they often get exactly what they pay for. For example, in May 1998, the Codex Food Labeling Committee—part of the international organization that sets food standards (under the rules of GATT)—rejected a consumer-led call for mandatory labeling of genetically altered foods. The ruling came despite consumer surveys in North America, Europe, and Australia consistently showing that an overwhelming majority—ranging from 72 percent in Germany to 89 percent in Australia and 93 percent in the U.S.—want genetically engineered foods to be labeled as such. The Codex did agree to propose some definitions of genetically engineered foods and to propose the labeling of engineered foods containing known allergens. Delegations from India and Norway made strong statements supporting consumers' right to information, as did observers from Greenpeace International, the Center for Science in the Public Interest, and organic farmers. However, delegations from the United States, Canada, New Zealand, and Australia—representing the interests of the industry—recommended much more limited labeling requirements.

Wisdom Traditions and the Meaning of Life

In the short run, the kind of activism engaged in by the Foundation on Economic Trends, the Alliance for Bio-Integrity, Greenpeace, the Rural Advancement Foundation International (RAFI), the Union of Concerned Scientists, the International Center for Technology Assessment, the Council for Responsible Genetics, and other public-interest groups is essential if we are to slow and control the implementation of genetic technologies. In the long run, however, we need more than activism. Our society's adoption of new technologies must be guided by something other than the profit motive; but what, precisely, will provide this guiding role?

Since our civilization's passionate embrace of new technologies seems rooted in religious visions of transcendence and transformation, it's likely that only a more meaningful and satisfying vision will be able to supplant it. Our collective response to biotechnology—if we wish to prevent the toxic tragedy of the industrial revolution from reaching the cellular and molecular levels—must therefore spring from a new biological ethic, even a new spirituality. Whether we couch it in scientific or theological terms, our response to biotech must be essentially spiritual, one that sees life as intrinsically meaningful and that proposes a set of human priorities in which profit does not dominate.

When we look to our religious heritage for clues from whence such a new spirituality might come, we find a somewhat confusing message. While some ancient religious teachings enjoin us to respect all of life, others seem to encourage a utilitarian view of nature.

In his influential 1967 essay, "The Historical Roots of Our Ecological Crisis," Lynn White, Jr., wrote:

> Christianity inherited from Judaism not only a concept of time as nonrepetitive and linear but also a striking story of creation. By gradual stages a loving and all-powerful God had created light and darkness, the heavenly bodies, the earth and all its plants, animals, birds, and fishes. Finally, God had created Adam and, as an afterthought, Eve to keep man from being lonely. Man named all the animals, thus establishing his dominance over them. God planned all of this explicitly for man's benefit and rule: no item in the physical creation had any purpose save to serve man's purposes. And, although man's body is made of clay, he is not simply part of nature: he is made in God's image.
>
> Especially in its Western form, Christianity is the most anthropocentric religion the world has seen. As early

> as the 2nd century both Tertullian and Saint Irenaeus of Lyons were insisting that when God shaped Adam he was foreshadowing the image of the incarnate Christ, the Second Adam. Man shares, in great measure, God's transcendence of nature. Christianity, in absolute contrast to ancient paganism and Asia's religions (except, perhaps, Zoroastrianism), not only established a dualism of man and nature but also insisted that it is God's will that man exploit nature for his proper ends.[11]

The anthropocentric orientation of the Hebraic tradition, later reinforced by the Greek Stoic view of human superiority over nature and by the Gnostic Christian belief in the corruptness of the physical world, culminated in the Baconian doctrine (discussed in chapter 5) of humanity's mandate to exploit and transform every aspect of an inherently valueless nature.

Islam, drawing upon Judaic and Christian texts, also partook of their otherworldly and human-centered world view. Muslim philosophers described nature as a hierarchical ladder of being with God at the top, followed by the angels, humanity, animals, plants, and minerals. According to the Qur'an, God created man as his *khalifa*—or deputy—on earth, whose duty is to look after creation and be good. God created the animals and plants—even the sun and moon—for humanity's benefit: "He constrained the night and day to be of service unto you." While Mohammed counseled love and benevolence among people, he did not extend this sphere of generosity to include animals, which are for food and other uses and are slaughtered for the two main feasts of the Muslim calendar.

Still, throughout other religions—and as a minor current within the Judeo-Christian and Islamic traditions—we find a very different trend of thinking. The utilitarian view of nature as "other"—a realm separate from the self, having value only to the degree that it serves human purposes—has always coexisted with the view of nature as an extension of the self or as the ground from which the self arises, which is hence imbued with *intrinsic* value and meaning.

One finds this thread woven prominently through the myths and rites of Native American and other tribal cultures. The following statement by Chief Segwalise of the Haudenosaunee clearly expresses an ecocentric world view:

> In our language, the Earth is our Mother Earth, the sun our Eldest Brother, the moon our Grandmother and so on. It is the belief of our people that all elements of the Natural World were created for the benefit of all living things and that we, as humans, are one of the weakest of the whole Creation, since we are totally dependent on the whole Creation for our survival.[12]

In China, the ancient Taoist sages taught a philosophy of gentle peacefulness based on emulating nature's way. Implicit in this philosophy was a critique of humans' attempts to bend nature to their will and the insight that technology is never entirely value free:

A small country has fewer people.
Though there are machines that can work ten to a hundred times faster than man, they are not needed.
The people take death seriously and do not travel far.
Though they have boats and carriages, no one uses them.
Though they have armor and weapons, no one displays them.
Men return to the knotting of rope in place of writing.
Their food is plain and good, their clothes fine but simple, their homes secure;
They are happy in their ways.[13]

Lao Tzu and Chuang Tzu (the primary ancient Taoist sources) continually repudiated the desire for wealth and power and advised people to let nature take its course. Nature is inherently wise, they said, and we should learn from it rather than seeking to dominate it for short-term gain. The more we try to control the world, the more chaotic it will become.

Throughout ancient Buddhist literature we find the teaching that all beings are sentient and contain the seed of the Buddha-nature. As one's mind progresses toward enlightenment, it also becomes increasingly aware of the essential unity of all animate life. Expanding awareness of the suffering in the world eventually leads to *mahakaruna,* a great compassion in which the individual identifies with all living beings.

While Hinduism represents an extremely diverse collection of beliefs and rites, throughout its ancient literature runs a thread of respect for life and for the awesome and mysterious universe.

According to the "Hymn to Goddess Earth" in the *Artharvaveda*:

> Thy snowy mountain heights, and thy forests, O earth, shall be kind to us! The brown, the black, the red, the multi-colored, the firm earth, that is protected by Indra, I have settled upon, not suppressed, slain, not wounded. . . .
>
> Rock, stone, dust, is this earth; this earth is supported, held together. To this golden-breasted earth I have rendered reverence.[14]

Even though Judaism, Christianity, and Islam present us with an anthropocentric view of the world, in these religions we still find at least an undercurrent of respect for the value within nature beyond its usefulness to humans. According to Louis Ginzberg in his authoritative *Legends of the Jews,* there is the sense in Judaism that there is nothing truly inanimate in nature and that "all things in creation are endowed with sensation."[15]

Jewish dietary rules were originally instituted as solutions to health problems and also as expressions of concern for the welfare of other sentient life. Rabbi Harold White explained it to me this way: "From a Jewish point of view, as we read the Bible, Adam and Eve are vegans. They are given herbs and lichens to eat; they're not allowed to eat meat. The first individual who is allowed to eat meat is Noah, and that's a concession. In the Jewish view of the messianic era, we will again return to vegetarianism and a harmonious relationship with nature. Even when the eating of meat was allowed, the only animals that you are allowed to eat, and the only fish and birds that you are allowed to eat, are ones that are vegetarians. You can only eat animals that chew their cuds; you can only eat fish that have fins and scales; you can only eat birds that have a crop in their neck as a grinding device for berries. This is to teach you that you should be noncarnivorous as well.

"You can't speak of the dietary laws in Judaism," continued White, "without seeing them as the final crystallization of concerns about humane treatment of animals. Take, for example, the injunction against mixing meat and milk. The ancient Hittites used to have a festival in the spring where they would boil a kid alive in its mother's milk. Judaism reacted against that and said, 'No, you can't mix meat and milk because that is a reminder of those sadistic practices.'"[16]

According to Sultain A. Ismail, Department of Zoology at The New College in Madras, India, "Islam means, a 'state of health or of nature'. . . The whole realm of nature is the revelation of the will of God. It is evident from the verses of the Holy Qur'an that the divine will is manifest in the creation of heavens and of earth, the alternation of day and night, and in the variety of plants and animals. There is intrinsic goodness, beauty, harmony, and orderliness in creation. . . . Those who live in peace and harmony with nature, making it instrumental to and congruous with God's moral law and purpose for mankind, are coworkers with God."[17] This sense of the oneness of the whole creation is referred to in the Qur'an (VI:38):

> There is not an animal (that lives) on the earth
> nor a being that flies on its wings,
> but (forms part of) the communities, like you.[18]

Muslims may turn to the strand of teaching in their tradition that sees nature as the revelation of the divine. As Dhu'n-Nun, an Egyptian Sufi mystic of the ninth century, put it, "I never hearken to the voices of the beasts or the rustle of the trees, the splashing of the waters or the song of the birds, the whistling of the wind or the rumble of the thunder, but I sense in them a testimony to Thy Unity. . . ."[19] Christians may recall Jesus' injunction to "consider the lilies how they grow: they toil not, they spin not; and yet I say unto you, that Solomon in all his glory was not arrayed like one of these"; or St. Francis's teaching that all living things are part of the fellowship of God.

Donald Conroy, Catholic theologian and ecologist, relates: "There is a phrase at the beginning of Genesis in the Hebrew scriptures that calls us to be stewards of the earth and caretakers of creation. To care for creation, to care for the garden, means to cultivate it and to replenish it. That comes even before the Ten Commandments, and it has great meaning for us today. It has implications for what we do, what we buy, how we act—both individually and as members of institutions. We see the results of not paying attention to that injunction in global warming, in the loss of diversity of species, and in the pollution of water and air. And so we have to take a responsible look at these things, and that has consequences as we begin to look at technological change."

Thus wherever we encounter some form or fragment of the perennial philosophy we find a variation on the message that while the sustenance of life requires predation—the killing and eating at least of plants—and hence *to some degree* a utilitarian attitude toward the other, nevertheless the basic thrust of spirituality is to steer us ever toward reverence and compassion and toward seeing the other as a subject like ourselves, not merely an object.

The perennial philosophy values empathy above efficiency. None of us would treat our children or pets as "efficiently" as possible, giving them the minimum of love in exchange for the maximum of obedience. Within a healthy family, it is understood that love is to be given for its own sake, without expectation of any specific result. The thrust of spiritual ethics is toward regarding all of creation as a family bound more by love than utility.

By itself, this implicit message of reverence and generosity toward life—the bottom line, if you will, of spiritual bioethics—might make us wary of biotechnology. It might also lead us to adopt a very different attitude toward nature than we in the modern industrial world have become accustomed to.

But in addition, the perennial philosophy also teaches an economic ethic. In *every* world religion—and prominently so in Judaism, Christianity, and Islam—we find injunctions against the accumulation of wealth for its own sake and praise for a materially simple lifestyle. Spiritual teachers tirelessly seek to turn us away from greed and toward generosity. Jesus said, "Blessed are the *poor*"—not "Blessed are the *rich*." Mohammed taught that usury—the charging of interest on loaned money—is a sin. And Tashunka Witko (Crazy Horse) said, "One does not sell the earth upon which people walk." Teachings like these, taken seriously, will surely evoke in us a philosophy about the world that precludes the patenting of genes and life forms and that leads us to question the commodification of the human body—as well as the concentration of economic power in corporations—that biotech relentlessly promotes.

While the critique of wealth and privilege is a universal constant in humanity's spiritual heritage, the specter of corporate monopoly over the basic essences of life has raised remarkably little concern among many people who regard themselves as spiritual or religious. Why aren't religious people morally outraged about corporate biotech? Why haven't they organized more pickets and boycotts? The reason has partly to do with recent history.

The twentieth-century contest between capitalism and communism has presented a confusing set of choices for spiritual philosophers. On one hand, capitalism openly promotes the concentration of wealth and the power that wealth brings—a clear contradiction of nearly all spiritual teachings. On the other hand, communism—which promises to distribute fairly the fruits of social production—eschews religion and offers a purely materialist interpretation of history. In post-World War II America, capitalism became identified with patriotism; in the East, communism became identified first with Leninism, then Stalinism. Most religious people understandably opposed the latter and became, if only by default, champions of the capitalist cause. Now, with the Cold War a fading memory, capitalism faces little organized resistance—either from socialists or religious groups—and the path appears to be open to the buying and selling of virtually everything on the planet, from forests to pollution rights to genes.

Perhaps it is time for people who regard themselves as spiritual or religious to go back to the teachings from which they originally drew inspiration. Beyond trying to be good, holy, enlightened, or saved, there is work to be done in *this* world—the created world of rocks, oceans, eagles, bears, redwood trees, and human communities. Spiritual seekers need the moral courage to say, in effect, "No! I will not sit mute, while greed consumes this planet. I will not by silence endorse the organized pillage of life on earth!"

Visions for the Future

There are already signs of the emergence of a new—or renewed—spirituality that holds sacred the complex web of living beings on our fragile planet. These signs are evident, for example, in the Deep Ecology movement, in the revival of nature-based indigenous religions, and in the formation of environmental awareness groups within mainstream denominations.

Suppose for a moment that these tender shoots were to flourish. Suppose we used the present ecological crisis as a wake-up call to transform our relationship with the natural world, learning once again to value diversity and mystery, and giving up our addiction to controlling and exploiting every aspect of the planet. How might society change? How might we then make use of our burgeoning scientific knowledge about genetics?

The details may be difficult to imagine. We've spent centuries building a society based on the manipulation of life, and the transition to a different way of being in the world—at once new and ancient—will take time. However, it is clear from the outset that a life-centered culture would emphasize local production for local consumption, using technologies that are human-scaled and that are understood and controlled by their users. Rather than directly manipulating genes in order to alter domesticated plant species or to improve human health, scientists would pay attention to the context of biological and social relationships that governs the expression of genes. Medical researchers would focus on environment, behavior, and lifestyle. Agriculturists, rather than relying on fewer and fewer hybrid or gene-engineered monocrops, would seek to preserve and reclaim the genetic heritage of thousands and millions of years, growing a wide diversity of locally adapted varieties using soil-preserving organic farming and gardening methods. Public policymakers would concentrate on taking care of *all* society's children rather than engineering perfect babies for a few wealthy parents.

How to get there from here? I see no obvious shortcuts. In the short term, more anticorporate activism is essential. But in the long term, we need nothing less than a cultural reorientation.

CHAPTER EIGHT

Cloning the Buddha?

Suppose your ten-year-old son were dying of a disease that could be cured only with a genetically engineered medicine. Would you then consider it morally wrong to prevent the development, or withhold the use, of that medicine?

Suppose you were a farmer living in a country facing famine and you had access to a genetically altered seed stock that was 30 percent more productive than your existing seed stocks. Would it be morally wrong to refuse the bioengineered seeds?

DURING THIS PAST year of intensive readings and discussions about biotechnology, I've gradually come to a conclusion that I had not initially expected to reach. At the outset, I anticipated that my views would eventually coalesce around the notion that some applications of biotech are morally good while others are bad and that we need a detailed public debate to figure out which is which. I thought we should consider a wide range of complex, hypothetical problems like the samples above and try to find nuanced answers that balance perceived benefits against known risks.

However, it is now quite clear that the development and

implementation of the technology is moving much faster than the public discussion. Momentous collective moral decisions are being made by default. To a large extent (especially in the case of the genetic engineering of foods), public debate is being deliberately foreclosed by powerful corporate interests.

I've also come to believe that, while it is possible to imagine a wide range of possible benefits from biotech, it is also possible to imagine risks on a scale that might far outweigh *any* conceivable benefit.

I realize that some bioethicists—people who have been contemplating these moral issues longer than I have and on a professional basis—have reached different conclusions. Frankly, in reading their arguments, I've found them often to be shallow or poorly informed. The ethicists appear naive about the nature of technological change—the fact that the adoption of a profoundly new technology is a social choice, not just an individual one, and that its costs are seldom obvious early in the process. Their desire to balance immediate, known benefits against immediate, known risks ignores the fact that many technological risks are neither immediate nor obvious. Risks are often initially subtle yet cumulative and at times ultimately overwhelming (as is the case with chemical-based industrial agriculture, the consequences of which are just now being realized in the forms of catastrophic topsoil loss, environmental pollution, and human health consequences ranging from lowered sperm counts to soaring incidences of chemical sensitivities).

The arguments now being advanced in favor of biotechnology exactly parallel those put forward for chemical pesticides, herbicides, and fertilizers in the 1940s and for nuclear energy in the 1950s. Now, as then, corporations tout benefits and downplay risks. Now, as then, the public is assured that the known benefits outweigh the unknown risks and that if there *are* problems, they can always be fixed easily later on.

In the cases of industrial agriculture and nuclear power, the solutions have not come easily. Agriculture has become the single most ecologically destructive activity on the planet. And we are drowning in radioactive waste: several tons are generated each day in this country alone, and we've run out of places to store the stuff.

In retrospect, what we really needed in the 1940s—instead of petrochemical-based industrial agriculture—was a refinement of existing agricultural techniques to strengthen the family farm, preserve topsoils, and make the food production process more diversified and resilient. And what we really needed in the 1950s—instead of nuclear power—was energy conservation. If we had decided to delay implementation of the new technologies for fifty years and spent that time studying their possible effects and investigating alternatives, what would we have lost? What would we have gained? In my view—and that of every ecologist I know—the gains likely would have far outweighed the losses.

Today we have a similar choice to make, and this time we ought to take the conservative approach. We need to slow drastically the process of technological change.

Might that mean that somewhere a child will die for lack of a treatment—or perhaps dozens or even hundreds of children? Maybe. Might it mean that farmers would have to forego planting seeds that, in some instances, could offer higher yields? Yes. But it might also mean that even more significant harms would be prevented. The ratio of costs to benefits simply isn't yet clear, and until it is we need to exercise great care. In the words of Margaret Mellon, "What it comes down to is that the risk-benefit equation is a much more difficult one than many people admit. You can usually articulate the benefits pretty well, but getting to an understanding of what the risks really are is very challenging." In the case of ecological risks, "We just don't know that much about how things interact in the environment. There's simply no way to do enough experiments to replicate all of the environmental interactions possible. So the risks may be much greater than we have thought. It's a difficult problem, made even more difficult by the fact that we tend to pour all of our scientific resources into the development of molecular biology, and we have comparatively little devoted to understanding soil ecology, for example."

This morning, as I sat down to write, I began (as I usually do) by checking my E-mail, and found a forwarded report about some work just published in the *Proceedings of the National Academy of Sciences*. It concerns a genetic parasite called a "group I intron," which can splice itself in and out of the genome of mitochondria (the little powerhouses of the cell that oxidize food in order to turn it into a form of energy that can be used for all living processes). The scientists (headed by Dr. Jeffrey Palmer) involved in the research have apparently found that this genetic parasite, which usually appears only in yeast, has recently been transferred to thousands of kinds of higher plants. "This massive wave of lateral transfers is of entirely recent occurrence," say the authors of the study, "perhaps triggered by some key shift in the intron's invasiveness within angiosperms [i.e., higher plants]."[1] The explosive invasion of the DNA sequence in question could have come about through the actions of viruses, insects, or bacteria. What it means is that environmental gene transfer can happen far more quickly and pervasively than most scientists had previously imagined. In their report, the researchers themselves raise concerns about releasing transgenic crops into the environment—especially as their finding comes on the heels of a separate report in the prestigious journal *Nature* suggesting that genes engineered into transgenic plants can be up to 30 times more likely to escape than the plant's own genes.[2]

One can't help but wonder whether this massive horizontal gene transfer from yeast to higher plants may actually have been triggered by commercial genetic engineering, since the procedure makes use of artificial genetic parasites as vectors in order to transfer genes between unrelated species.

Horizontal gene transfer is risky business indeed. Suppose, through genetic engineering, we inadvertently transfer herbicide resistance to a host of weeds. Similar kinds of explosive horizontal gene transfer have already been documented among viruses and bacteria that are developing antibiotic resistance.

Can the risks entailed in biotechnology be overstated? Of course; for example, the *Boys from Brazil* scenario of mad Nazi scientists whipping up batches of Hitler clones is far fetched. But the real, demonstrable risks are certainly horrific enough. In a paper entitled "Redesigning the World: Ethical Questions about

Genetic Engineering," Buddhist scholar Ron Epstein writes, "It now seems likely, unless a major shift in international policy occurs quickly, that the major ecosystems that support the biosphere are going to be irreversibly disrupted, and genetically engineered viruses may very well lead to the eventual demise of almost all human life. In the course of the major transformations that are on the way, it also seems likely that human beings will be transformed, both intentionally and unintentionally, in ways that will make us something different from what we now consider human."[3]

Clearly, neither Epstein nor anyone else can prove that "genetically engineered viruses" will "lead to the demise of almost all human life." That the possibility exists, however, is denied by no one I know of who is familiar with the present state of the research. After all, in addition to the billions of dollars currently spent developing genetically engineered seeds and medicines, several national governments are also spending hundreds of millions to create designer-germ weapons of mass destruction that have exactly this capability.

When I first read Epstein's shocking forecast, I thought it was overstated and calculated to frighten. But after rereading it repeatedly and analyzing it closely, I find that I cannot argue with any specific assertion it contains. The risks really are that apocalyptic.

Would *every* existing or proposed application of biotechnology lead toward such dire results? Obviously not. But the task of separating which ones might and which might not is complex and difficult and one which our society has hardly begun to undertake. Thus would it really be immoral to abort the development of a disease cure or a new seed variety in order to forestall—or at least further assess—these potential risks? I think that just the opposite is true—it would be unethical for us *not* to cease our headlong rush toward the implementation of biotechnology, even if doing so means foregoing some immediate apparent benefits.

Beyond Risks and Benefits

Risk-benefit analysis is essential in making moral and ethical decisions, but it is not sufficient in and of itself. That's because, as we try to add up all the known risks and balance them against all the known benefits, we sometimes forget about ethical considerations that are difficult to quantify or even to verbalize.

I started thinking more along these lines after my conversation with Margaret Mellon. When I asked about her own moral qualms regarding biotech, she replied, "I do see moral implications. But I think some folks don't like biotechnology—I'm actually in that category—and for them, one way of attacking it is to say that it is morally or ethically reprehensible and then to throw everything but the kitchen sink into that bin. I think some things belong in that bin and some things don't. I think we need to think harder than we have about what a moral framework is. What is it that it is wrong to do regarding the environment or other people? What are the moral principles that are at work here? It certainly is wrong to kill or injure other people. Is it wrong to kill a bacterium? Is it wrong to kill a plant? What is the principle that would underlie the assessment of the moral impact of the technology? I don't think folks have thought as hard as they really should about that issue. For some, what is immoral is simply 'anything I don't like.' This technology does touch on moral issues, but I think we need to know what those are, and they need to be unearthed and articulated rather than assumed and implied."

Mellon's comments have stuck with me. Repeatedly I've asked myself, Am I mindlessly throwing everything having to do with biotech into the "unethical" bin? I've tried not to; yet the more I've learned about biotechnology, the more skeptical I've become. And much of my skepticism hinges on exactly the moral issues that Mellon alluded to—ones that do indeed require unearthing. Although I believe there are probably many such issues, at this point I'm able to define only three. Still, these by themselves amount to powerful and sweeping reasons for caution:

1) Does nature have inherent meaning or only utilitarian value? That is, do blue jays and maple trees have some fundamental right to exist and express themselves autonomously *apart from* whatever benefit they may offer us (even simply by way of their beauty)—or is their existence of interest to us only to the extent that we can derive some profit from it? If we decide that the former is the case, then we should pause before embarking on the wholesale implementation of agricultural biotechnology (with either seeds or farm animals), because once we start down that path there is nothing to prevent *all life* either from becom-

ing raw material for the humanly organized process of production or from being permanently affected by genetic pollution.

2) Is it morally wrong to create or exacerbate inequalities within a society? Government and businesses sometimes take actions that heighten existing social or economic inequalities—such as changing welfare laws or shifting manufacturing operations to low-wage countries. Some people believe such actions are morally wrong; others disagree.

The use of germ-line gene therapy and cloning within an already economically stratified nation is almost sure to lead to a *genetic* stratification of that nation. Genetic discrimination in hiring, promotions, and the issuing of health insurance policies is likely to become commonplace, even if legally discouraged. Over time, whatever inequalities already exist in the society will become ever more genetically justified, fixed, and pronounced. If equality of opportunity is morally desirable, then the introduction of human genetic engineering techniques within an already inequitable society is bound to raise profound new ethical dilemmas.

3) Given a centralized industrial society in which nearly all food is grown and distributed by way of private corporations, should those corporations be required to label foods so as to allow citizens to choose intelligently what they wish to eat? If we see democracy as having a moral value, then the failure to label genetically altered foods has distinctly moral implications. Genetically altered foods may or may not pose risks—I believe the

evidence suggests that they do—but this is a problem quite separate from the question of whether people should have the right to know what they're eating.

These are not matters of risk assessment. Intelligent, responsible people may differ in their approach to these issues, but the issues themselves cannot be ignored if we are to arrive at a moral assessment of biotechnology; they *must be* articulated and openly discussed.

What to Do?

Throughout these months of research and writing, I've maintained a focus on developing practical recommendations. If biotechnology is indeed as morally problematic as the evidence I've cited suggests, then what should be done?

In my early thinking, I tended toward caution. After all, who am I to condemn whole industries? It would be easy enough to make sweeping recommendations for the elimination of genetic technologies, but (assuming those recommendations were followed) I personally wouldn't have to live with the consequences in the same way that a research scientist would—or a farmer, a medical technician, or a patient with a genetic disease. If I am to critically examine the moral impacts of biotechnology, I must be prepared to do the same with regard to any remedy I propose.

I was emboldened in this regard by my conversation with ecologist Philip Regal. When asked what he thought should be done, he said, "You can't imagine how much time I've put into this. I've been to scores of government workshops and I've organized several of the important ones. I've studied the issues; I've written papers. I've put so much of my life into trying to get them to do this safely, and I now think that they just won't. I don't think I wasted all those years, but I think I might have smelled the coffee a little earlier.

"For fifteen years I've said that we can cut down the risks if we have more research and we train people who are competent regulators, and if we have responsible agencies that know how to deal with these scientifically complex issues. I've held out all kinds of hope that we could do that one day, because I do see a lot of potential benefits that could come from biotechnology. I wanted to see it proceed, but proceed safely.

"If we had spoken a month ago I might have answered differently, but you're getting me on a day when I'm starting to think that it's hopeless and we ought to just have a moratorium. On the food issue, it's been clear that the industry is just going to fight any effective regulation. On the environmental issue, it seems that they're going to fight regulation. Politically, the world has not done anything to handle this in a way that gives me any comfort whatsoever. . . .

"As far as medical research goes, I know that there are stunning things going on with genetic engineering as a research tool. Biotechnology is sensational; it's just wonderful. I get excited about what it can do in terms of basic research. But on the other hand, I really am concerned about its misuse, whether it's genetically engineering people or cloning people.

"Maybe it's time for me to give up my pleas for more

research and regulation. Maybe I've held out too long. The decision to just let market forces develop it is an insane one."

A moratorium. Some form of moratorium on further implementation of medical biotechnology, and a complete temporary ban on agricultural biotech, seem essential. But how to produce these results?

Laws appear necessary. As blunt an instrument as legislation may be, I can think of no other tool to achieve something like a moratorium. But what kinds of laws?

I'm a firm believer in the principle that the ends we achieve always reflect the means we use; if we wish to achieve moral ends, our means must also be inherently moral. In the present instance, this would require that we *not* use inherently unfair legal restrictions in order to achieve a morally desirable consequence. If, for example, Congress were to pass a law making genetic research a federal crime, that would be an excessively repressive and thus inherently immoral action—even if it had the "good" effect of putting the brakes on human cloning and agricultural biotech.

Are there any kinds of legislation having the effect of drastically slowing the implementation of biotechnology that would be *inherently* fair and moral? I would offer two concrete suggestions:

- legislation rescinding the applicability of patent law to any organism or DNA sequence. It was a legal and moral mistake to allow the patenting of life forms and genes to begin with.

- legislation or policy ruling requiring the clear labeling of any food or food ingredient that is genetically altered. This should be done in any case; apart from the goal of restricting the technology, this is simply a matter of protecting the democratic rights of citizens.

Achieving these legislative goals will require citizen activism. The second will be the easier of the two: ultimately all that's necessary is a change in FDA policy in the U.S. (and in the equivalent regulatory agencies in other countries). It is possible that a lawsuit recently filed by the Center for Technology Assessment will succeed in forcing this policy change in the U.S.; if the suit is not successful, citizens must demand that their representatives draft and pass a law requiring labeling.

The first of these suggestions is more problematic. The banning of life patents will be resisted bitterly by the biotech industry, which has many friends in the U.S. Congress. Its implementation must be the goal of an intensive, sustained, global activist campaign.

If you wish to have your voice heard in favor of labeling genetically engineered foods, or for banning patents on life forms, contact one of the activist groups listed in the Resources section of this book.

Is it possible that, in the context of a different society—one with ecologically and spiritually enlightened priorities—*some* of the techniques of genetic engineering might be beneficially employed?

Probably so. But the fact remains that our own society is so exploitative and so inequitable that genuinely helpful uses of biotech are likely to be overwhelmed by disastrous ones. And so vigorous efforts to rein in the gene corporations (and I believe that both kinds of legislation outlined above would have precisely that effect) are desperately needed.

As we've seen, one cannot morally assess any new technology without examining the entire framework of the society from which the technology is emerging. What motives are driving the technology's development? Who will benefit? Who will be hurt? Who makes the decisions?

Ultimately, if we as a society wish to employ some form of genetic technology for truly beneficial purposes we must begin, not with the technology itself, but with an ethical reappraisal and reform of our collective institutions and priorities.

I'm sure that those who stand to benefit economically from biotech will not find this advice appealing or perhaps even sensible. It amounts to a call for a new moral, spiritual, ecological consciousness—one with profound implications not just for biotech but for our economic and political systems as well.

The fact is, in order to handle a technology as powerful as genetic engineering, we humans need to be far more compassionate and wise than we currently give evidence of being. What's the best way to move in that direction? *That's* the question we should be asking ourselves.

I believe that, in the end, the best way to clone the Buddha—to create a society of caring, responsible, creative, joyful people—will *not* be by manipulating genes, but rather by working diligently on our own personal moral refinement, collectively confronting power and its abuses, and creating a nurturing context for our children and our grandchildren. In a way, there's nothing new about that prescription: it's one with which the Buddha himself—along with Jesus, Amos, Mohammed, and Lao Tzu—would almost certainly have agreed.

One can only pray that we're at last ready to take it to heart.

APPENDIX

State of the World Forum Statement on Life and Evolution

LIFE IS AN intimate web of relations that evolves in its own right, interfacing and integrating its myriad diverse elements. The complexity and interdependence of all forms of life have the consequence that the process of evolution cannot be controlled, though it can be influenced. It involves an unpredictable creative unfolding that calls for sensitive participation from all the players, particularly from the youngest, most recent arrivals: human beings.

Life must not be treated as a commodity that can be owned, in whole or in part, by anyone, including those who wish to manipulate it in order to design new life forms for human convenience and profit. There should be no patents on organisms or their parts. We must also recognize the potential dangers of genetic engineering to health and biodiversity and the ethical problems it poses for our responsibilities to life. We propose a moratorium on commercial releases of genetically engineered products and a comprehensive public enquiry into the legitimate and safe uses of genetic engineering. This enquiry should take account of the precautionary principle as a criterion of sensitive participation in living processes. Species should be respected for their intrinsic natures and valued for their unique qualities, on which the whole intricate network of life depends.

We recognize the validity of the different ways of knowing that have been developed in different cultures and the equivalent value of the knowledge gained within these traditions. These add substantially to the set of alternative technologies that can be used for the sustainable use of natural resources that will allow us to preserve the diversity of species and to pass the precious gift of life in all its beauty and creativity to our children and their children, to the next century and beyond.

Signatories include:

Fritjof Capra, Institute for Ecoliteracy, California, U.S.A.
Peter Fenwick, Scientific and Medical Network, U.K.
Brian Goodwin, Schumacher College, U.K.
Mae-Wan Ho, Open University, U.K.
Ervin Laszlo, Club of Budapest, Hungary
David Lorimer, Scientific and Medical Network, U.K.
Richard Strohman, University of California, Berkeley, U.S.A.
Marilyn Schlitz, Institute of Noetic Sciences, California, U.S.A.
Peter Saunders, King's College, London, U.K.

Notes

INTRODUCTION

1. Erwin Chargaff, *Heraclitean Fire: Sketches from a Life Before Nature,* quoted in *Natural Law,* accessed 14 October 1998, <www.natural-law.ca/genetic/ScientistsonDangers.html>.
2. See the Resources section, pages 253–256.
3. Lee Silver, *Remaking Eden: Cloning and Beyond in a Brave New World* (New York: Avon, 1997), 235.
4. Examples of these efforts include Richard B. Brandt, *A Theory of the Good and Right* (Amherst, N.Y.: Prometheus, 1997); and Joseph L. Daleiden, *The Science of Morality: The Individual, Community, and Future Generations* (Amherst, N.Y.: Prometheus, 1998).
5. Prince Charles Philip Arthur George, "Seeds of Disaster," *Living Earth,* no. 199, July–September 1998, 6.
6. Aldous Huxley, *The Perennial Philosophy* (New York: Harper & Row, 1944).

CHAPTER ONE

1. Lynn Margulis and Dorion Sagan, *Slanted Truths: Essays on Gaia, Symbiosis, and Evolution* (New York: Copernicus, 1997), 272.
2. Ibid., 273.

3. Michael Denton, *Evolution: A Theory in Crisis* (Bethesda, Md.: Adler & Adler, 1986), 149–51.

4. Richard Tapper, "Changing Messages in the Genes," *New Scientist*, 25 March 1989.

5. Quoted in Craig Holdrege, *Genetics & the Manipulation of Life: The Forgotten Factor of Context* (Hudson, N.Y.: Lindisfarne Press, 1996), 78 (emphasis added).

6. Philip Regal, telephone conversation with the author, 20 October 1998.

7. Holdrege, *Genetics & the Manipulation of Life*, 79

8. Mae-Wan Ho, *Genetic Engineering—Dream or Nightmare: The Brave New World of Bad Science and Big Business* (Bath, England: Gateway, 1998), 109.

9. Lily Kay, *The Molecular Vision of Life: Caltech, the Rockefeller Foundation, and the Rise of the New Biology* (New York: Oxford University Press, 1993), 43.

10. Ibid., 3.

11. Ibid., 6.

12. Dean Hamer, *Living With Our Genes: Why They Matter More than You Think* (New York: Doubleday, 1998), 6.

13. Quoted in Robert Wesson, *Beyond Natural Selection* (Cambridge, Mass.: MIT Press, 1993), 279.

14. See Ho, *Genetic Engineering*, 195–7; and Wesson, *Beyond Natural Selection*, 278–83.

15. Margaret Mellon, telephone conversation with the author, 5 November 1998.

16. Wesson, *Beyond Natural Selection*, 82–3.

17. Quoted in ibid., 281.

18. Ho, *Genetic Engineering*, 100.

19. Wesson, *Beyond Natural Selection*, 283.

20. Quoted in ibid., 234.

21. Ibid.

22. Anne S. Moffatt, "A Challenge to Evolutionary Biology, *American Scientist* 77, May/June, 1989, 224.

23. Fritjof Capra, *The Web of Life: A New Understanding of Living Systems* (New York: Anchor, 1996), 227.

24. Wesson, *Beyond Natural Selection*, 291.

25. See Rupert Sheldrake, *A New Science of Life: The Hypothesis of Formative Causation* (Los Angeles: Tarcher, 1981), 49.

26. Ibid., 98.

27. Ibid., 122.

28. Rupert Sheldrake, *Seven Experiments That Could Change the World: A Do-It Yourself Guide to Revolutionary Science* (New York: Riverhead, 1995).

29. Capra, *Web of Life*, 176.

30. Ibid., 227–8.

31. Quoted in ibid., 267.

32. Quoted in Wesson, *Beyond Natural Selection*, 279.

33. Freeman Dyson, "Mankind's Place in the Cosmos," *U. S. News and World Report*, 18 April 1998, 72.

34. Margulis and Sagan, *Slanted Truths*, 279.

CHAPTER TWO

1. Lynn Margulis and Dorion Sagan, *Slanted Truths: Essays on Gaia, Symbiosis, and Evolution* (New York: Copernicus, 1997), 271–2.

2. Ibid., 279.

3. Jeremy Rifkin, *Algeny: A New Word—A New World* (New York: Penguin, 1984), 208.

4. For more discussion of the rise of mechanistic philosophy and its dominance of biology, see Jeremy Rifkin, *Algeny*; and *The Biotech Century: Harnessing the Gene and Remaking the World* (New York: Tarcher/Putnam, 1998).

5. Mae-Wan Ho, *Genetic Engineering—Dream or Nightmare: The Brave New World of Bad Science and Big Business* (Bath, England: Gateway, 1998), 62.

6. Robert Augros and George Stanciu, *The New Biology: Discovering the Wisdom in Nature* (Boston: Shambhala, 1987), 196.

7. Ibid., 197.

8. Ibid., 197–8.

9. For further discussion on remote viewing, see Russell Targ and Jane Katra, *Miracles of Mind: Exploring the Healing Powers of Non-local Consciousness* (Novato, Calif.: New World Library, 1998).

10. Rupert Sheldrake, *Seven Experiments That Could Change the World: A Do-It Yourself Guide to Revolutionary Science* (New York: Riverhead, 1995), 9–96.

11. Paul Davies, *God and the New Physics* (New York: Simon & Schuster, 1983), 8.

12. Rupert Sheldrake, *A New Science of Life: The Hypothesis of Formative Causation* (Los Angeles: Tarcher, 1981), 203.

13. Richard Dawkins has attempted to offer step-by-step explanations for the origin of complex systems through natural selection in books such as *The Blind Watchmaker* (London: W. W. Norton, 1985), but he has been refuted by Michael Behe in *Darwin's Black Box: The Biochemical Challenge to Evolution* (New York: Simon & Schuster, 1996) and by Robert Wesson in *Beyond Natural Selection* (Cambridge, Mass.: MIT Press, 1993).

14. Fritjof Capra, *The Web of Life: A New Understanding of Living Systems* (New York: Anchor, 1996), 7.

15. See discussions of cooperation in nature in Augros and Stanciu, *The New Biology*, 89–129; and Peter Kropotkin, *Mutual Aid: A Factor of Evolution* (London: Freedom Press, 1987), 21–73.

CHAPTER THREE

1. Quoted in Jeremy Rifkin, *Algeny: A New Word—A New World* (New York: Penguin, 1984), 132.

2. Quoted in ibid., 133.

3. Quoted in Gina Kolata, *Clone: The Road to Dolly and the Path Ahead* (New York: William Morrow, 1998), 33.

4. This version of the events was recounted by Philip Regal in a telephone conversation with the author, 20 October 1998.

5. Ibid.

6. Michael Pollan, "Playing God in the Garden," *New York Times Magazine*, 25 October 1998.

CHAPTER FOUR

1. Quoted in David Noble, *The Religion of Technology: The Divinity of Man and the Spirit of Invention* (New York: Knopf, 1998), 84.

2. Quoted in ibid., 109.

3. See Vandana Shiva, *Biopiracy: The Plunder of Nature and Knowledge* (Boston: South End Press, 1997), 13.

4. Quoted in Jeremy Rifkin, *The Biotech Century: Harnessing the Gene and Remaking the World* (New York: Tarcher/Putnam, 1998), 42.

5. Quoted in ibid., 42.

6. Quoted in Eric S. Grace, *Biotechnology Unzipped: Promises & Realities* (Washington, D.C.: Joseph Henry Press, 1997), 204–5.

7. See discussions of the NIH and gene patenting in Rifkin, *Biotech Century*, 58; and Grace, *Biotech Unzipped*, 200.

8. See Rifkin, *Biotech Century*, 60.

9. See ibid., 59–60.

10. See ibid., 63.

11. Ibid., 64.

12. Quoted in "Customer Profiles," *Pilot Software*, retrieved 12 May 1998, <www.pilotsw.com/synergy/profile/monsanto.htm>.

13. "Monsanto's Empire," *New Internationalist*, Lindsay Troub, ed., retrieved 12 May 1998, <www.oneworld.org/ni/issue293/monsanto.html>.

14. Ibid.

15. Monsanto Web site, accessed 12 May 1998, <www.monsanto.com>.

16. The following are examples of prominent regulatory officials who have moved back and forth between government programs or agencies and the biotech industry (as of September 1998):

    David W. Beier: former head of Government Affairs for Genentech, Inc.; now chief domestic policy advisor to Al Gore, Vice President of the United States

    Linda J. Fisher: former assistant administrator of the United States Environmental Protection Agency's Office of Pollution Prevention Pesticides and Toxic Substances; now Vice President of Government and Public Affairs for Monsanto Corporation

    L. Val Gidings: former biotechnology regulator and (biosafety) negotiator at the United States Department of Agriculture (USDA/APHIS); now Vice President for Food & Agriculture of the Biotechnology Industry Organization (BIO)

    Marcia Hale: former assistant to the President of the United States and Director for Intergovernmental Affairs; now Director of International Government Affairs for Monsanto Corporation

    Michael (Mickey) Kantor: former Secretary of the United States Department of Commerce and former Trade Representative of the United

State; now member of the board of directors of Monsanto Corporation

JOSH KING: former Director of Production for White House events; now Director of Global Communication in the Washington, D.C. office of Monsanto Corporation

TERRY MEDLEY: former administrator of the Animal and Plant Health Inspection Service (APHIS) of the United States Department of Agriculture, former chair and vice chair of the United States Department of Agriculture Biotechnology Council, and former member of the U.S. Food and Drug Administration (FDA)'s Food Advisory Committee; now Director of Regulatory and External Affairs of DuPont Corporation's Agricultural Enterprise

MARGARET MILLER: former chemical laboratory supervisor for Monsanto; now Deputy Director of Human Food Safety and Consultative Services, New Animal Drug Evaluation Office, Center for Veterinary Medicine in the United States Food and Drug Administration (FDA)

WILLIAM D. RUCKELSHAUS: former chief administrator of the United States Environmental Protection Agency (USEPA); now (and for the past 12 years) a member of the board of directors of Monsanto Corporation

MICHAEL TAYLOR: former legal advisor to the United States Food and Drug Administration (FDA)'s Bureau of Medical Devices and Bureau of Foods; later executive assistant to the Commissioner of the FDA; still later a partner at the law firm of King & Spaulding where he supervised a nine-lawyer group whose clients included Monsanto Agricultural Company; still later Deputy Commissioner for Policy at the United States Food and Drug Administration; and now again with the law firm of King & Spaulding

Lidia Watrud: former microbial biotechnology researcher at Monsanto Corporation in St. Louis, Missouri; now with the United States Environmental Protection Agency Environmental Effects Laboratory, Western Ecology Division

Clayton K. Yeutter: former Secretary of the U.S. Department of Agriculture, former U.S. Trade Representative (who led the U.S. team in negotiating the U.S. Canada Free Trade Agreement and helped launch the Uruguay Round of the GATT negotiations); now a member of the board of directors of Mycogen Corporation, whose majority owner is Dow AgroSciences, a wholly owned subsidiary of The Dow Chemical Company

(Data provided by The Edmonds Institute: International Center for Technology Assessment, 20319 92nd Avenue West, Edmonds, WA, 98020)

17. rBGH-free labeling is now allowed if accompanied by a disclaimer noting that the USDA has found no health risks associated with rBGH.

18. See Ronnie Cummins, "Monsanto's Genetically Engineered Products Meet Resistance," *AWI Quarterly*, Spring/Summer 1997, 10.

19. Cathy Donohue, "Biotechnology and the New Toxic Cotton," *Food & Water Journal*, Spring 1998, 36.

20. See Rifkin, *Biotech Century*, 84; and Josie Glausiusz, "The Great Gene Escape," *Discover*, May 1998, 96. While studies show that glyphosate (the active ingredient in Roundup) is less toxic than many other herbicide chemicals, it is nevertheless the third most commonly

reported cause of illness among Californian agricultural workers, and for landscape maintenance workers it ranks first. Glyphosate reduces the ability of bacteria to transform nitrogen into a usable form for plants, and it harms fungi that help plants absorb water and nutrients. Residues of the herbicide have been found in lettuce, carrots, and barley that were planted a year after the soil was sprayed. Critics contend that as farmers plant more Roundup Ready seeds and spray their fields with larger doses of Roundup, herbicide "drift" may increase significantly. If this happens, neighboring farms may be forced to switch to the Monsanto seeds in order to keep their crops from being destroyed by airborne glyphosate.

21. Quoted in Rural Advancement Foundation International (RAFI), 9 October 1998 news release, retrieved 9 October 1998, <www.rafi.org>.

22. Quoted by Leora Broydo in "A Seedy Business," *Mother Jones*, 7 April 1998.

23. David Letourneau, conversation with the author, Occidental, California, 10 October 1998.

24. Quoted in Rural Advancement Foundation International (RAFI), 1 November 1998 news release, retrieved 1 November 1998, <www.rafi.org>.

25. See Susan Benson and Leora Broydo, "Flavr Savd," *Mother Jones*, January/February, 1997.

26. Quoted in Ronnie Cummins, Organic Consumers Association, "Food Bytes" 31 October 1998, internet press release, retrieved 31 October 1998, <www.purefood.org>.

27. Quoted by Cathy Donohue, "Biotechnology and the New Cotton," *Food & Water Journal*, Spring 1998, 36.

28. See, for example, David Korten, *When Corporations Rule the World* (San Francisco: Kumarian Press and Barrett-Koehler, 1995). Relevant organizations include:

    The Program on Corporations, Law, and Democracy (POCLAD), P. O. Box 246, South Yarmouth, MA 02664, <people@poclad.org>

    The Alliance for Democracy, P. O. Box 683, Lincoln, MA 01773, <www.ea1.com/alliance>

    International Forum on Globalization (IFG), 1555 Pacific Avenue, San Francisco, CA 94109, <www.ifg.org>

CHAPTER FIVE

1. G. J. V. Nossal, *Reshaping Life* (New York: Cambridge University Press, 1985; reprinted in William Dudley, ed., *Genetic Engineering: Opposing Viewpoints*, San Diego: Greenhaven Press, 1990), 46–52 (page citations are to the reprint edition).

2. Jerry Mander, *Four Arguments for the Elimination of Television* (New York: William Morrow/Quill, 1977).

3. James Burke and Robert Ornstein, *The Axemaker's Gift: Technology's Capture and Control of Our Minds and Culture* (New York: Tarcher/Putnam, 1995), xvii.

4. Poll data obtained from The Council for Responsible Genetics, 5 Upland Road, Suite 3, Cambridge, MA 02140, (617) 868-0870, <www.essential.org/crg/>.

5. David Letourneau, conversation with the author, Occidental, California, 10 October 1998.

6. Arthur Caplan, "If Gene Therapy Is the Cure, What Is the Disease?" The Center for Bioethics, accessed 12 December 1998, <www.med.upenn.edu/~bioethic/>.

7. Ibid.

8. Ibid.

9. Ibid.

10. Gilbert Meilander, *Bioethics: A Primer for Christians* (Grand Rapids, Mich.: William B. Eerdmans Publishing Company, 1996), 42–3.

11. Dr. Donald Conroy, telephone conversation with the author, 16 November 1998.

12. Dr. John Reigstad, telephone conversation with the author, 19 October 1998.

13. In June 1998, Prince Charles of Britain wrote an article, published in the London *Daily Telegraph*, in which he decried biotechnology and urged more reliance on organic methods in agriculture. Genetic engineering, according to the article, represents a human attempt to tamper with God's realm. "We live in an age of rights," according to Charles. "It seems to me that it is time our Creator had some rights too."

14. Dr. Colin Gracey, telephone conversation with the author, 13 November 1998.

15. Quoted in Dorothy Nelkin and M. Susan Lindee, *The DNA Mystique: The Gene As a Cultural Icon* (New York: W. H. Freeman, 1995), 55.

16. Ron Epstein, "Ethical and Spiritual Issues in Genetic Engineering," *Ahimsa Voices*, vol. 5, no. 4, October 1998, 5.

17. Alliance for Bio-Integrity Web site, accessed 9 October 1998, <www.bio-integrity.org>.

18. John Fagan, *Genetic Engineering: The Hazards/Vedic Engineering: The Solutions* (Fairfield, Iowa: Maharishi University Press, 1995).

19. "Manipulating the Forms of Life," *World Goodwill Newsletter*, no. 4, 1997, 3.

20. Ron Epstein, conversation with the author, 23 October 1998.

21. Bruce R. Reichenbach and V. Elving Anderson, *On Behalf of God: A Christian Ethic for Biology* (Grand Rapids, Mich.: Eerdmans, 1995), 183.

22. Ibid., 196.

23. Ibid., 205.

24. Ibid., 214.

25. Philip Heffner, "Cloning As Quintessential Human Act," Chicago Center for Religion and Science, accessed 14 December 1998, <www.usao.edu>.

26. Ingrid Shafer, "Biotechnology: The Moral Challenge of Human Cloning," Chicago Center for Religion and Science, accessed 14 December 1998, <www.usao.edu>.

27. Ibid.

28. C. S. Lewis, *The Abolition of Man* (New York: Macmillan, 1947), 69, 71.

CHAPTER SIX

1. Dorothy Nelkin and M. Susan Lindee, *The DNA Mystique: The Gene As a Cultural Icon* (New York: W. H. Freeman, 1995), 2, 40.

2. Ibid., 40–1.

3. Ibid., 86.

4. Alistair McIntosh, "The Cult of Technology," *Resurgence,* no. 188, May/June 1998, 8–11.

5. David Noble, *The Religion of Technology: The Divinity of Man and the Spirit of Invention* (New York; Knopf, 1998), 3.

6. Ibid., 12.

7. Lewis Mumford, *The Pentagon of Power,* vol. 2 of *The Myth of the Machine* (New York: Harcourt Brace Jovanovich, 1970), 264.

8. Quoted in Noble, *The Religion of Technology*, 17.

9. Quoted in ibid., 49.

10. Quoted in ibid., 173–4, 183.

11. Jeremy Rifkin, *The Biotech Century: Harnessing the Gene and Remaking the World* (New York: Tarcher/Putnam, 1998), 118.

12. Quoted in Nelkin and Lindee, *DNA Mystique*, 23.

13. See Rifkin, *Biotech Century*, 123.

14. Quoted in John F. Kilner, Rebecca D. Pentz, and Frank E. Young, eds., *Genetic Ethics: Do the Ends Justify the Genes?* (Grand Rapids, Mich.: Eerdmans, 1997), 35.

15. See Nelkin and Lindee, *DNA Mystique*, 189.

16. Arthur Caplan, "If Gene Therapy Is the Cure, What Is the Disease?" The Center for Bioethics, accessed 12 December 1998, <www.med.upenn.edu/~bioethic/>.

17. Mumford, *The Pentagon of Power*, 197.

CHAPTER SEVEN

1. Erwin Chargaff, *Heraclitean Fire: Sketches from a Life Before Nature*, quoted in *Natural Law*, accessed 14 October 1998, <www.natural-law.ca/genetic/ScientistsonDangers.html>.

2. George Wald, "The Case Against Genetic Engineering," *The Recombinant DNA Debate*, David A. Jackson and Stephen Stich, eds. (Englewood Cliffs, N.J.: Prentice Hall, 1979), 127–8.

3. Quoted in *Alive: Canadian Journal of Health and Nutrition*, February 1998.

4. "New Plants Threaten Bees," *The Futurist*, May 1998, 13.

5. Michael W. Fox, "Genetic Engineering: Nature's Cornucopia or Pandora's Box?," *The Animals' Agenda*, March 1987, 11.

6. Quoted in Jeremy Rifkin, *Algeny: A New Word—A New World* (New York: Penguin, 1984), 32.

7. Monsanto Web site, accessed 12 May 1998, <www.monsanto.com>.

8. Frances Moore Lappé, Joseph Collins, and Peter Rosset with Luis Esparza, *World Hunger: Twelve Myths*, 2nd ed. (New York: Food First, 1998).

9. Andrew Kimbrell, "Why Biotechnology and High-Tech Agriculture Cannot Feed the World," *The Ecologist* 28, no. 5, Sepember/October, 1998, 294.

10. Ibid., 294–8.

11. Lynn White, "The Historical Roots of our Ecologic Crisis," *Science*, vol. 155, 10 March 1967, 1203–7.

12. Quoted in Peter Marshall, *Nature's Web: An Exploration of Ecological Thinking* (New York: Simon & Schuster, 1992), 142.

13. Quoted in ibid., 21.

14. Quoted in ibid., 44.

15. Quoted in ibid., 103.

16. Rabbi Harold White, telephone conversation with the author, 18 October 1998.

17. Sultain A. Ismail, "Environment: An Islamic Perspective," University of Oregon, accessed 8 September 1998, <http://darkwing.uoregon.edu/~jbonine/islamenviro.html>.

18. Quoted in ibid.

19. Quoted in Marshall, *Nature's Web*, 135.

CHAPTER EIGHT

1. Quoted in Mae-Won Ho, "Horizontal Gene Transfer—New Evidence," E-mail to the author, 4 December 1998.

2. Ibid.

3. Ron Epstein, "Redesigning the World: Ethical Questions about Genetic Engineering," accessed 26 December 1998, <http://online.sfsu.edu/~rone/GE%20Essays/Redesigning.html>.

Resources

ACTIVIST ORGANIZATIONS

The Center for Ethics and Toxics (CETOS). Newsletter: *Genetics/Toxics Stopwatch*. P. O. Box 673, Gualala, CA 95445. Phone: (707) 884-1700; E-mail: cetos@cetos.org; Web site: www.cetos.org

Council for Responsible Genetics (CRG). Newsletter: *Genewatch: Monitoring the Social Impact of Biotechnology*. Phone: (617) 868-0870; Fax: (617) 491-5344; E-mail: crg@essential.org

Environment Research Foundation. Newsletter: *Rachel's Environment & Health Weekly*. P. O. Box 5036, Annapolis, MD 21403. Fax: (410) 263-8944; E-mail: erf@rachel.org; Web site: www.monitor.net/rachel/

Mothers for Natural Law, American Campaign to Ban Genetically Engineered Foods, P. O. Box 1177, Fairfield, IA 52556. Phone: (515) 472-2809; Fax: (515) 472-2683; E-mail: mothers@lisco.com; Web site: www.lisco.com/mother

Campaign for Food Safety/Organic Consumers Association. 860 Hwy 61, Little Marais, MN 55614. Phone: (218) 226-4164; Fax (218) 226-4157; E-mail: alliance@mr.net; Web site: www.purefood.org

Rural Advancement Foundation International–USA (RAFI), 101 Hillsboro St., Rm. 5, P. O. Box 655, Pittsboro, NC 27312. Phone: (919) 542-1396; Fax: (919) 542-2460; E-mail: EcoNet@rafiusa; Web site: www.rafi.org

Union of Concerned Scientists. 1616 P Street N.W., Suite 310, Washington, DC 20036. Phone: (202) 332-0900; Fax: (202) 332-0905.

RECOMMENDED READINGS

Augros, Robert, and George Stanciu. *The New Biology: Discovering the Wisdom in Nature*. Boston: Shambhala, 1987.

Burke, James, and Robert Ornstein. *The Axemaker's Gift: Technology's Capture and Control of Our Minds and Culture*. New York: Tarcher/Putnam, 1995.

Capra, Fritjof. *The Web of Life: A New Understanding of Living Systems*. New York: Anchor, 1996.

Dudley, William, ed. *Genetic Engineering: Opposing Viewpoints*. San Diego: Greenhaven Press, 1990.

Fagan, John. *Genetic Engineering: The Hazards/Vedic Engineering: The Solutions*. Fairfield, Iowa: Maharishi University Press, 1995.

Fox, Michael W. *Superpigs and Wondercorn: The Brave New World of Biotechnology . . . And Where It All May Lead*. New York: Lyons & Burford, 1992.

Grace, Eric S. *Biotechnology Unzipped: Promises & Realities*. Washington, D.C.: Joseph Henry Press, 1997.

Ho, Mae-Wan. *Genetic Engineering—Dream or Nightmare: The Brave New World of Bad Science and Big Business*. Bath, England: Gateway, 1998.

Holdrege, Craig. *Genetics & the Manipulation of Life: The Forgotten Factor of Context*. Hudson, N.Y.: Lindisfarne Press, 1996.

Kay, Lily. *The Molecular Vision of Life: Caltech, the Rockefeller Foundation, and the Rise of the New Biology*. New York: Oxford University Press, 1993.

Kevles, Daniel J. *In the Name of Eugenics: Genetics and the Uses of Human Heredity*. New York: Knopf, 1985.

Kevles, Daniel J., and Leroy Hood, eds. *The Code of Codes: Scientific and Social Issues in the Human Genome Project.* Cambridge, Mass.: Harvard University Press, 1992.

Kimbrell, Andrew. *The Human Body Shop: The Engineering and Marketing of Life.* San Francisco: HarperSanFrancisco, 1993

Lappé, Marc. *Broken Code: The Exploitation of DNA.* San Francisco: Sierra Club, 1984.

Lappé, Marc, and Britt Bailey. *Against the Grain.* Boston: Common Courage Press, 1998.

Margulis, Lynn, and Dorion Sagan, *Slanted Truths: Essays on Gaia, Symbiosis, and Evolution.* New York: Copernicus, 1997.

Nelkin, Dorothy, and M. Susan Lindee, *The DNA Mystique: The Gene As a Cultural Icon.* New York: W. H. Freeman, 1995.

Noble, David. *The Religion of Technology: The Divinity of Man and the Spirit of Invention.* New York: Knopf, 1998.

Rifkin, Jeremy. *The Biotech Century: Harnessing the Gene and Remaking the World.* New York: Tarcher/Putnam, 1998.

Shiva, Vandana. *Biopiracy: The Plunder of Nature and Knowledge.* Boston: South End Press, 1997.

Suzuki, David, and Peter Knudtson. *Genethics: The Clash Between the New Genetics and Human Values.* Cambridge, Mass.: Harvard University Press, 1989; 2d ed., 1998.

Wesson, Robert. *Beyond Natural Selection.* Cambridge, Mass.: MIT Press, 1993.

Yoxen, Edward. *The Gene Business: Who Should Control Biotechnology?* New York: Harper & Row, 1983.

Index

BIOGRAPHIES

Dorion Sagan and Lynn Margulis

Foreword Writers of
Cloning the Buddha:
The Moral Impact of Biotechnology
Richard Heinberg

DORION SAGAN is a writer primarily in the fields of evolutionary biology, philosophy, and history of science. His works include essays and books such as "Metametazoa: Biology and Multiplicity" (1992), "Partial Closure" (1998), and *Biospheres* (1990). Among the several books he has coauthored with Lynn Margulis, *Mystery Dance: On the Evolution of Human Sexuality* (1991) has been translated into seven languages. Dorion Sagan's current projects include works of fiction as well as *Into the Cool* with Eric D. Schneider, a popularization of nonequilibrium thermodynamics. He resides in Northampton, Massachusetts with his son, Tonio.

LYNN MARGULIS, PH.D., is internationally known for her research on cell biology and the evolution of small forms of life. She helped James E. Lovelock develop the "Gaia Hypothesis," which states that the Earth is a single self-regulating system with its own physiology.

Spanning a wide range of scientific topics, Lynn Margulis's many authored or coauthored publications range from professional monographs to children's literature. Her books with Dorion Sagan include *What Is Life?* (1995), *What Is Sex?* (1998), and

Microcosmos: Four Billion Years of Microbial Evolution (1990). Her book *Symbiosis in Cell Evolution* (1993, second edition) is considered a classic of 20th century biology.

Dr. Margulis is an elected member of the U.S. National Academy of Sciences (1983), the World Academy of Art and Science (1995), the Russian Academy of Natural Sciences (1997), and the American Academy of Arts and Sciences (1998). She is former chairman of the National Academy of Science's Space Science Board Committee on Planetary Biology and Chemical Evolution. She received a NASA Public Service Award in 1981 and codirects NASA's Planetary Biology Internship (PBI) program, administered through the Marine Biological Laboratory in Woods Hole, Massachusetts.

A Guggenheim Fellow, Dr. Margulis has been awarded seven honorary doctorates. She is a Distinguished University Professor in the Department of Geosciences at the University of Massachusetts in Amherst, where she resides.

QUEST BOOKS
are published by
The Theosophical Society in America
Wheaton, Illinois 60189-0270,
a branch of a world organization
dedicated to the promotion of the unity of humanity and the encouragement of the study of religion, philosophy, and science to the end that we may better understand ourselves and our place in the universe. The Society stands for complete freedom of individual search and belief.
For further information about its activities,
write, call 1-800-669-1571, or consult its Web page:
http://www.theosophical.org

*The Theosophical Publishing House
is aided by the generous support of
THE KERN FOUNDATION,
a trust established by Herbert A. Kern
and dedicated to Theosophical education.*